Food Packaging: Principles and Practices

Food Packaging: Principles and Practices

Renate Herrera

CLANRYE
INTERNATIONAL
www.clanryeinternational.com

Clanrye International,
750 Third Avenue, 9th Floor,
New York, NY 10017, USA

ISBN: 978-1-63240-898-3

Cataloging-in-Publication Data

Food packaging : principles and practices / Renate Herrera.
p. cm.
Includes bibliographical references and index.
ISBN 978-1-63240-898-3
1. Food--Packaging. 2. Food--Packaging--Technological innovations. 3. Food industry and trade--Appropriate technology. I. Herrera, Renate.
TP374 .F66 2019
664.09--dc23

For information on all Clanrye International publications visit our website at www.clanryeinternational.com

CLANRYE INTERNATIONAL

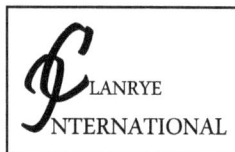

Contents

Preface

Food is packaged for protection against tampering and spoilage, efficient handling and for ease of distribution. It is also meant to present information regarding its nutritional profile and sale specifications and thus encourage marketing. Food packaging is an essential aspect of the food industry. Food can be packaged into boxes, cartons, cans, trays, wrappers and pallets. A diverse range of machinery is used for food packaging such as check weighing machines, cartoning machines, conveying and accumulating machines, filling machines, vacuum-packaging machines, etc. This textbook provides comprehensive insights into food packaging. Most of the topics introduced herein cover the principles and practices of food packaging. This book, with its detailed analyses and data, will prove immensely beneficial to professionals and students involved in this area at various levels.

To facilitate a deeper understanding of the contents of this book a short introduction of every chapter is written below:

Chapter 1- The packaging of food is a measure for protection, tamper resistance or for providing nutritional information, etc. This chapter has been carefully written to provide an introduction to food packaging, through the inclusion of the varied aspects of packaging development, packaging environments, role of food packaging and functions environments grid.

Chapter 2- Shelf life of food is the longest time duration that food can be stored without becoming unfit for consumption. An understanding of the shelf life of food requires a detailed study of the processes of food deterioration and food additives that can be used for improving the durability of food. This chapter covers all the diverse aspects which affect the shelf life of food such as food deterioration, food additives, etc.

Chapter 3- The material used for food packaging directly impacts its durability for consumption. This chapter closely examines the different materials used for food packaging, such as metal packaging materials, glass packaging materials, plastic packaging materials, and edible and bio based packaging materials, for a comprehensive understanding of food packaging.

Chapter 4- There are diverse machineries that are used for food packaging. Some of these are multihead weigher, sealing machines, case sealer, cartoning machine, vacuum-packaging machines, wrapping machines, etc. which have been carefully examined in this chapter.

Chapter 5- To improve the shelf life of food products, the composition of the internal atmosphere of a package is modified. There may be a reduction in oxygen or replacement of it with different gases. The concentration of gases such as carbon dioxide and nitrogen may also be controlled. This chapter discusses in elaborate details the crucial aspects of controlled and modified atmospheric packaging, such as gas flush, foodstuffs in MAP and gas technology for modified atmosphere packaging.

Chapter 6- Package testing ideally involves the examination of the package contents, levels of packaging, packaging materials, etc. through various qualitative and quantitative procedures. The topics covered in this chapter address the different types of package testing, such as thickness testing, moisture vapor transmission rate, oxygen transmission rate, oxygen scavenger and carbon dioxide transmission rate for a complete understanding.

Chapter 7- Package design is an integral part of the food packaging process. It involves the identification of the requirements of marketing, shelf-life, structural design, logistics, graphic design, etc. This chapter discusses in extensive detail the different types of package design and the basics of milk package design.

Chapter 8- Food labels include critical information pertaining to the contents of the food package, its ingredients or information regarding certain allergy risks such as presence of soy or gluten. This chapter gives an analytical view on food labels, labeling policies and government regulations for a comprehensive understanding of the industrial standards of food packaging and labeling.

Chapter 9- Food packaging is aimed at protecting food between the stages of processing and consumption. After consumption, the food packaging needs to be disposed in a responsible manner. It is a major contributor to municipal solid waste (MSW).The use of plastic and non-biodegradable materials for packaging creates negative impact on the environment on disposal.

Chapter 10- The food packaging industry has witnessed significant innovations in packaging materials in recent years. Modern food packaging strives to be higher, durable, sustainable, renewable and biobased. This chapter will explore the upcoming trends in food packaging such as Liqui Glide, Amcor's Liqui FormTM bottle, Atomic Layer Deposition (ALD), etc.

I would like to share the credit of this book with my editorial team who worked tirelessly on this book. I owe the completion of this book to the never-ending support of my family, who supported me throughout the project.

Renate Herrera

Introduction to Food Packaging

The packaging of food is a measure for protection, tamper resistance or for providing nutritional information, etc. This chapter has been carefully written to provide an introduction to food packaging, through the inclusion of the varied aspects of packaging development, packaging environments, role of food packaging and functions environments grid.

In today's society, packaging is pervasive and essential. It surrounds, enhances and protects the goods we buy, from processing and manufacturing through handling and storage to the final consumer. Without packaging, materials handling would be a messy, inefficient and costly exercise, and modern consumer marketing would be virtually impossible.

Packaging lies at the very heart of the modern industry, and successful packaging technologists must bring to their professional duties a wide-ranging background drawn from a multitude of disciplines. Efficient packaging is a necessity for almost every type of product whether it is mined, grown, hunted, extracted or manufactured. It is an essential link between the product makers and their customers. Unless the packaging operation is performed correctly, the reputation of the product will suffer and the goodwill of the customer will be lost. All the skill, quality and reliability built into the product during development and production will be wasted, unless care is taken to see that it reaches the user in the correct condition. Properly designed packaging is the main way of ensuring safe delivery to the final user in good condition at an economical cost.

Functions of Packaging

Packaging has four primary functions i.e. containment, protection, convenience and communication.

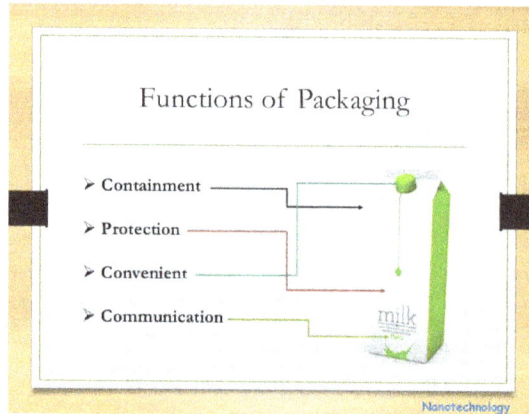

Functions of Packaging
- Containment
- Protection
- Convenient
- Communication

Nanotechnology

Containment

All products must be contained before they can be moved from one place to another. The "package", whether it is a bottle of cola or a bulk cement rail wagon, must contain the product to function successfully. Without containment, product loss and pollution would be wide spread.

The containment function of packaging makes a huge contribution to protecting the environment from the myriad of products which are moved from one place to another. Faulty packaging (or under packaging) could result in major pollution of the environment.

Protection

This is often regarded as the primary function of the package: to protect its contents from outside environmental effects, such as water, moisture vapour, gases, odours, micro-organisms, dust, shocks, vibrations and compressive forces, and to protect the environment from the product.

For the majority of food products, the protection afforded by the package is an essential part of the preservation process. For example, aseptically packaged milk and fruit juices in paperboard cartons only remain aseptic for as long as the package provides protection. Likewise, vacuum packaged meat will not achieve its desired shelf life if the package permits oxygen to enter. In general, once the integrity of the package is breached, the product is no longer preserved.

Packaging also protects or conserves much of the energy expended during the production and processing of the product. For example, to produce, transport, sell and store 1 kg of bread requires 15.8 MJ (mega joules) of energy. This energy is required in the form of transport fuel, heat, power and refrigeration in farming and milling the wheat, baking and retailing the bread and distributing both the raw materials and the finished product. To produce the low density polyethylene (LDPE) bag to package a 1 kg loaf of bread requires 1.4 MJ of energy. This means that each unit of energy in the packaging protects 11 units of energy in the product. While eliminating the packaging might save 1.4 MJ of energy, it would also lead to spoilage of the bread and a consequent waste of 15.8 MJ of energy.

Convenience

Trend towards "grazing" (i.e., eating snack type meals frequently and on-the run, rather than regular meals), the demand for a wide variety of food and drink at outdoor functions such as sports events and leisure time, have created a demand for greater convenience in household products. The products designed around principles of convenience include foods which are pre-prepared and can be cooked or reheated in a very short time, preferably without removing them from their primary package. Sauces, dressings and condiments that can be applied simply through aerosol or pump-action packages minimize mess. Thus packaging plays an important role in meeting the demands of consumers for convenience.

Two other aspects of convenience are important in package design. One of these can best be described as the apportionment function of packaging. In this context, the package functions by reducing the output from industrial production to a manageable, desirable "consumer" size. Thus, a vat of wine is "apportioned" into bottles, a churn of butter is "apportioned" by packing into 25 ml packet and a batch of ice cream is "apportioned" into 2 L plastic tubs.

An associated aspect is the shape (relative proportions) of the primary package with regard to consumer convenience (Ex., easy to hold, open and pour as appropriate) and efficiency in building into secondary and tertiary packages. In the movement of packaged goods in interstate and international trade, it is clearly inefficient to handle each primary package individually. Here, packaging plays another very important role in permitting primary packages to be unitized into secondary packages (Ex., placed inside a corrugated case) and secondary packages to be unitized into a tertiary package (Ex., a stretch-wrapped pallet). This unitizing acitivity can be carried a stage further to produce a quarternary package (Ex., A container which is loaded with several pallets). As a consequence of this unitizing function, handling is optimized since only a minimal number of discrete packages or loads need to be handled.

Communication

A package functions as a "silent salesman". The modem methods of consumer marketing would fail were it not for the messages communicated by the package. The ability of consumers to instantly recognize products through distinctive branding and labeling enables supermarkets to function on a self-service basis. Without this communication function (i.e., if there were only plain packs and standard package sizes), the weekly shopping expedition to the supermarket would become a lengthy, frustrating nightmare as consumers attempted to make purchasing decisions without the numerous clues provided by the graphics and the distinctive shapes of the packaging.

PACKAGINGDEVELOPMENT

Food packaging is defined as a coordinated system of preparing food for transport, distribution, storage, retailing, and end-use to satisfy the ultimate consumer with optimal cost. Food packaging is an essential part of modern society; commercially processed food could not be handled and distributed safely and efficiently without packaging. The World Packaging Organization estimates that more than 25% of food is wasted because of poor packaging. Thus, it is clear that optimal packaging can reduce the large amount of food waste. Moreover, the current consumer demand for convenient and high-quality food products has increased the impact of food packaging.

We can categorize packaging systems into four groups: primary packaging, secondary packaging, distribution or tertiary packaging, and unit load.

- Primary packaging

  The first-level package that directly contacts the product is referred to as the "primary package." For example, a beverage can or a jar, a paper envelope for a tea bag, an inner bag in a cereal box, and an individual candy wrap in a pouch are primary packages, and their main function is to contain and preserve the product. Primary packages must be non-toxic and compatible with the food and should not cause any changes in food attributes such as color changes, undesired chemical reactions, flavor, etc.

- Secondary packaging

  The secondary package contains two or more primary packages and protects the primary packages from damage during shipment and storage. Secondary packages are also used to prevent dirt and contaminants from soiling the primary packages; they also unitize groups of primary packages. A shrink-wrap and a plastic ring connector that bundles two or more cans together to enhance ease of handling are examples of secondary packages.

- Tertiary package

  The tertiary package is the shipping container, which typically contains a number of the primary or secondary packages. It is also referred to as the "distribution package." A corrugated box is by far the most common form of tertiary package. Its main function is to protect the product during distribution and to provide for efficient handling.

- Unit load

  A unit load means a group of tertiary packages assembled into a single unit. If the corrugated boxes are placed on a pallet and stretch wrapped for mechanical handling, shipping and storage, the single unit is referred to as a "unit load." The objective is to aid in the automated handling of larger amounts of product. A fork-lift truck or similar equipment is used to transport the unit load.

Role of Food Packaging

The principal roles of food packaging are to protect food products from outside influences and damage, to contain the food, and to provide consumers with ingredient and nutritional information (Coles 2003). Traceability, convenience, and tamper indication are secondary functions of increasing importance. The goal of food packaging is to contain food in a cost-effective way that satisfies industry requirements and consumer desires, maintains food safety, and minimizes environmental impact.

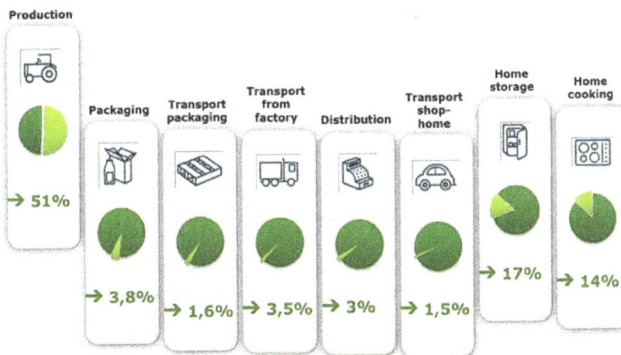

Protection/Preservation

Food packaging can retard product deterioration, retain the beneficial effects of processing, extend shelf life, and maintain or increase the quality and safety of food. In doing so, packaging provides protection from 3 major classes of external influences: chemical, biological, and physical.

Chemical protection minimizes compositional changes triggered by environmental influences such as exposure to gases (typically oxygen), moisture (gain or loss), or light (visible, infrared, or ultraviolet). Many different packaging materials can provide a chemical barrier. Glass and metals provide a nearly absolute barrier to chemical and other environmental agents, but few packages are purely glass or metal since closure devices are added to facilitate both filling and emptying. Closure devices may contain materials that allow minimal levels of permeability. For example, plastic caps have some permeability to gases and vapors, as do the gasket materials used in caps to facilitate closure and in metal can lids to allow sealing after filling. Plastic packaging offers a large range of barrier properties but is generally more permeable than glass or metal.

Biological protection provides a barrier to microorganisms (pathogens and spoiling agents), insects, rodents, and other animals, thereby preventing disease and spoilage. In addition, biological barriers maintain conditions to control senescence (ripening and aging). Such barriers function via a multiplicity of mechanisms, including preventing access to the product, preventing odor transmission, and maintaining the internal environment of the package.

Physical protection shields food from mechanical damage and includes cushioning against the shock and vibration encountered during distribution. Typically developed from paperboard and corrugated materials, physical barriers resist impacts, abrasions, and crushing damage, so they are widely used as shipping containers and as packaging for delicate foods such as eggs and fresh fruits. Appropriate physical packaging also protects consumers from various hazards. For example, child-resistant closures hinder access to potentially dangerous products. In addition, the substitution of plastic packaging for products ranging from shampoo to soda bottles has reduced the danger from broken glass containers.

Containment and Food Waste Reduction

Any assessment of food packaging's impact on the environment must consider the positive benefits of reduced food waste throughout the supply chain. Significant food wastage has been reported in many countries, ranging from 25% for food grain to 50% for fruits and vegetables (FAO 1989). Inadequate preservation/protection, storage, and transportation have been cited as causes of food waste. Packaging reduces total waste by extending the shelf life of foods, thereby prolonging their usability. Rathje and others (1985) found that the per capita waste generated in Mexico City contained less packaging, more food waste, and one-third more total waste than generated in comparable U.S. cities. In addition, Rathje and others (1985) observed that packaged foods result in 2.5% total waste—as compared to 50% for fresh foods—in part because agricultural by-products collected at the processing plant are used for other purposes while those generated at home are typically discarded. Therefore, packaging may contribute to the reduction of total solid waste.

Marketing and Information

A package is the face of a product and often is the only product exposure consumers experience prior to purchase. Consequently, distinctive or innovative packaging can boost sales in a competitive environment. The package may be designed to enhance the product image and/or to differentiate the product from the competition. For example, larger labels may be used to accommodate recipes. Packaging also provides information to the consumer. For example, package labeling satisfies legal requirements for product identification, nutritional value, ingredient declaration, net weight, and manufacturer information. Additionally, the package conveys important information about the product such as cooking instructions, brand identification, and pricing. All of these enhancements may impact waste disposal.

Traceability

The Codex Alimentarius Commission defines traceability as "the ability to follow the movement of a food through specified stage(s) of production, processing and distribution". Traceability has 3 objectives: to improve supply management, to facilitate traceback for food safety and quality purposes, and to differentiate and market foods with subtle or undetectable quality attributes. Food manufacturing companies incorporate unique codes onto the package labels of their products; this allows them to track their products throughout the distribution process. Codes are available in various formats (for example, printed barcodes or electronic radio frequency identification [RFID]) and can be read manually and/or by machine.

Convenience

Convenience features such as ease of access, handling, and disposal; product visibility; resealability; and microwavability greatly influence package innovation. As a consequence, packaging plays a vital role in minimizing the effort necessary to prepare and serve foods. Oven-safe trays, boil-in bags, and microwavable packaging enable consumers to cook an entire meal with virtually no preparation. New closure designs supply ease of opening, resealability, and special dispensing features. For example, a cookie manufacturer recently introduced a flexible bag with a scored section that provides access to the cookies. A membrane with a peelable seal covers the opening before sale and allows reclosure after opening. Advances in food packaging have facilitated the development of modern retail formats that offer consumers the convenience of 1-stop shopping and the availability of food from around the world. These convenience features add value and competitive advantages to products but may also influence the amount and type of packaging waste requiring disposal.

Tamper Indication

Willful tampering with food and pharmaceutical products has resulted in special packaging features designed to reduce or eliminate the risk of tampering and adulter-ation. Although any package can be breeched, tamper-evident features cannot easily be replaced. Tamper-evident features include banding, special membranes, breakaway closures, and special printing on bottle liners or composite cans such as graphics or text that irreversibly change upon opening. Special printing also includes holograms that cannot be easily duplicated. Tamper-evident packaging usually requires addition-al packaging materials, which exacerbates disposal issues, but the benefits generally outweigh any drawback. An example of a tamper-evident feature that requires no addi-tional packaging materials is a heat seal used on medical packaging that is chemically formulated to change color when opened.

Other Functions

Packaging may serve other functions, such as a carrier for premiums (for example, inclusion of a gift, additional product, or coupon) or containers for household use. The potential for packaging use/reuse eliminates or delays entry to the waste stream.

Chemicals in Food Packaging

Per-and polyfluoroalkyl substances, or PFAS, are a family of greaseproof, waterproof and nonstick industrial compounds. They're used in hundreds of consumer products, including ones that touch your food. These chemicals pollute the bodies of almost ev-eryone worldwide, and have been linked to a slew of serious health problems.

Some of the most worrisome places these chemicals lurk are in fast food wrappers and takeout containers. Food and Drug Administration tests found that PFAS chemicals can migrate out of food wrappers to contaminate food, especially when the food is greasy. And when EWG and colleagues tested fast food wrappers, we found fluorinated chemicals in 40 percent of the wrappers tested. This included packaging for sandwich-es, pizza, fried chicken and pastries.

Until companies change their packaging, or laws are put in place to keep our food safe from this nasty class of chemicals, PFAS in fast food packages is one more reason to cut back on fast food and greasy carryout whenever possible. Avoiding these substances may be even more important if you are pregnant or have kids, as PFAS chemicals can be particularly harmful to a developing fetus or young child.

Babies and young children are exposed to these chemicals in more ways than adults. They can ingest PFAS chemicals by drinking breast milk, crawling on dusty floors, and putting their hands in their mouths after touching contaminated materials. Because of their small size, children may have higher exposures by body weight than adults.

Toxic fluorinated chemicals can lower a baby's birth weight when the mother is exposed. Women drinking water contaminated with the PFAS chemical PFOA in West Virginia and Ohio had increased risk of pregnancy-induced hypertension and pre-eclampsia. PFAS chemicals at concentrations common in Americans may reduce the effectiveness of vaccines in children.

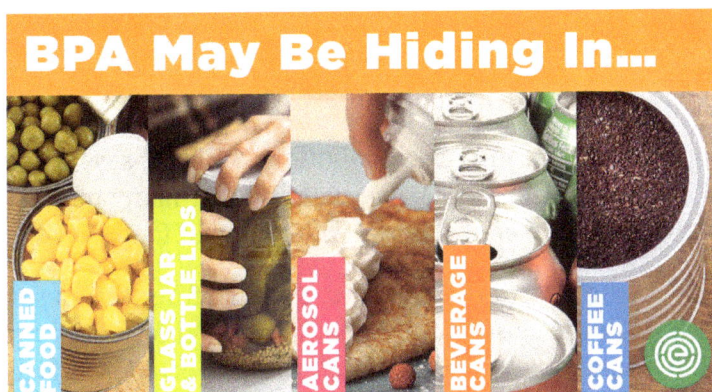

Researchers from the Silent Spring Institute; the University of California, Berkeley; the University of Notre Dame; and other institutions collected more than 400 samples of packaging from fast-food restaurants across the country—including Chipotle, McDonald's, and Subway—and found that 33 percent of them contained some form of the chemical fluorine.

"We've all heard that eating more fresh foods is better for our health for a wide range of reasons," says Laurel Schaider, Ph.D., research scientist with the Silent Spring Institute and the lead study author. "This study provides another reason why."

The good news, however, is that most fast-food packaging did not contain any fluorine, Shaider says. This shows that some manufacturers might be using fluorine compound-free chemicals to get the water- and grease-resistant effects they want without using compounds that carry a health risk, she says.

Not all of the chemicals in the fluorine family are harmful. However, experts are most concerned with a group of fluorinated chemicals called per- and polyfluoroalkyl substances, or PFASs. PFASs are often used to coat food wrappers and containers so that grease and water don't leak through them.

But they've been linked to health problems, such as some cancers, decreased fertility, hypertension in pregnancy, low birth weight, thyroid disease, and a weakened immune system. These chemicals stay in the body (and in the environment) for long periods of time.

Although this study didn't test food items for PFASs, other studies have shown that these chemicals can leach from the packaging to the food, especially when the food is hot.

When the researchers analyzed a subset of 20 pieces of fast-food packaging, they found PFASs in all of them, and six contained perfluorooctanoic acid (PFOA). PFOA is one of the fluorine compounds that have been most strongly linked to health problems.

Schaider says this study shows that fluorine can be a useful marker for the presence of harmful PFASs.

"The fact that they're so prevalent is worrisome," Schaider says. "Millions of Americans, including children, eat fast food every day."

It's especially concerning that the researchers found PFOA, says Consumer Reports' senior scientist Michael Hansen, Ph.D., because in 2011, many manufacturers agreed to voluntarily stop using the chemical in the production of food packaging.

And even when PFOA isn't used, Hansen says, manufacturers often substitute similar fluorinated chemicals in the PFAS family that aren't as thoroughly studied as PFOA. Preliminary research suggests that these substitutes also could be linked to some of the same dangerous health effects.

Lynn Dyer, president of the Foodservice Packaging Institute, the trade association for the North American food packaging industry, says that all chemicals used in food packaging "go through rigorous testing to ensure that they meet stringent U.S. Food and Drug Administration regulations, providing the safe delivery of foods and beverages to consumers".

Reducing the Exposure

Aside from cutting down on fast food altogether, Hansen says, one thing consumers can do is to limit the amount of time they leave your food in its packaging. If you can, once you arrive home or at the office, take food out of wrappers and use your own plates and bowls instead.

You might also want to consider what type of packaging your food is delivered in. The researchers found that more than half of dessert and bread wrappers and 38 percent of burger and sandwich wrappers they tested in their study contained fluorine. Just 20 percent of paperboard, which might hold French fries, was found to have fluorine.

Packaging Development

Early Packaging

Examples of early packaging include flex woven baskets, leaves, gourds, wooden barrel, pottery containers and glass. Glass packaging has a very long history but early types of glass packaging were very expensive and very rare form of packaging.

Figure: A modern example of early types of packaging using bamboo,
straw and paper used to package mangos

Nineteenth Century Packaging Development

The 19th Century from 1800 – 1900 was a period of rapid development of packaging systems with the appearance of:

- Metal cans (1818)

- Paper bag (1850s)

- Folding paperboard carton (1880s)

- Corrugated paperboard case (1890s)

- Tubes (eg toothpaste tubes)

- Milk Bottles (1860s).

Fruit and vegetables continue for the most part to be unpackaged and if any was shipped they would be in wooden boxes. The use of wooden boxes for packaging and shipping horticultural products continued until the end of the 20th century. Wooden boxes were phased out in NZ with the development of strong and cheap corrugated cases and then finally by the use of reusable plastic crates.

Package Environments

The packaging has to perform its functions in three different environments. Failure to consider all three environments during package development will result in poorly designed packages, increased costs, consumer complaints and even avoidance or rejection of the product by the customer.

Physical Environment

This is the environment in which physical damage can be caused to the product. It includes shocks from drops, falls and bumps; damage from vibration arising from transportation modes including road, rail, sea and air; and compression and crushing damage arising from stacking in warehouses and during transportation, or in the home environment.

Ambient Environment

This is the environment, which surrounds the package. Damage to the product can be caused as a result of gases (particularly oxygen), water and water vapor, light (particularly UV radiation), and the effects of heat and cold, as well as micro- and macro-organisms which are ubiquitous in many warehouses and retail outlets. Contaminants in the ambient environment such as exhaust fumes from automobiles and dust and dirt can also find their way into the product unless the package acts as an effective barrier.

Human Environment

This is the environment in which the package interacts with people, and designing packages for this environment requires knowledge of the vision and strength capabilities and limitations of humans, as well as legislative and regulatory requirements. Since one of the functions of the package is to communicate, it is important that the messages are received clearly by consumers. In addition, the package must contain information required by law such as product description and net weight.

To maximize its convenience functions, the package should be simple to hold, open and use by the consumer. For a product, which is not totally consumed when the package is first opened, the package should be able to be resealed and retain the quality of the product until completely used. Furthermore, the package should contain a portion size, which is also convenient for the intended consumers; a package, which contained too much product that deteriorated before being completely consumed clearly contains too large a portion.

Functions/Environments Grid

Figure: Functions/environments grid for evaluating package performance

The functions of packaging and the environments where the package has to perform can be laid out in a two-way matrix or grid as shown in figure above. Anything that is done in packaging can be classified and located in one or more of the 12-function/ environment cells. The grid provides a methodical yet simple way of evaluating the suitability of a particular package design before it is actually adopted and put into use. As well, the grid serves as a useful aid when evaluating existing packaging.

References

- Food-packaging, scientific-status-summaries, science-reports, read-ift-publications, knowledge-center: ift.org, Retrieved 19 April 2018

- These-toxic-chemicals-food-packaging-are-getting-your-meals-22091: ewg.org, Retrieved 13 April 2018

- Harmful-chemicals-in-fast-food-packaging: consumerreports.org, Retrieved 23 March 2018

- Evolution-food-packaging: begreenpackagingstore.com, Retrieved 11 July 2018

- Food-packaging-makul: nuristianah.lecture.ub.ac.id, Retrieved 30 May 2018

Shelf Life of Food

Shelf life of food is the longest time duration that food can be stored without becoming unfit for consumption. An understanding of the shelf life of food requires a detailed study of the processes of food deterioration and food additives that can be used for improving the durability of food. This chapter covers all the diverse aspects which affect the shelf life of food such as food deterioration, food additives, etc.

We understand shelf life to be the time during which a food maintains characteristics and a level of quality that is suitable for human consumption. In the food industry, the shelf life of a food is the time between the production or packaging of the product and the time when it becomes unacceptable under certain environmental conditions and when the consumption of said food implies a risk to consumer health.

There are several factors involved in the deterioration or loss of the original quality of a food. These factors can be divided into two types: intrinsic (inherent to the nature of the food itself) or extrinsic (external conditions facing food), and are determined by different quality parameters: organoleptic, nutritional, hygienic, physical, chemical or microbiological.

The intrinsic factors that affect shelf life are those that respond to the formulation of the food. In the food industry, it is imperative that the manufacturer has the following knowledge about its products:

- Raw materials

- Composition and formulation of the product (additives used)

- Water activity

- Total acidity and pH value

- Potential Redox

- Available oxygen

Taking all this information into account, the producer can choose the systems that maximize the life of a product according to the needs that it may have. For example, the oxidation of edible oils is a significant problem for the food industry due to the considerable increase in the use of fat and polyunsaturated oils (Frankel, 2010), so it is important to know the nutritional quality and the possible processes that the different raw materials have gone though, and to determine what antioxidants can slow down the oxidation process.

The extrinsic factors that affect the shelf life of the food are those that are present in the process, packaging and storage of the product. Mainly they are:

- Exposure to sunlight

- Temperature

- Humidity

- Damage to packaging

- Distribution and places of sale

During the different manipulation processes of the product, it is necessary to control its interaction with the components of the external system. To control of the process used every detail counts: the light permeability of the packaging, the distribution of humidity and the relative temperature, both in storage and in transportation, are the main external factors to be monitored and optimized.

Methodologies to Determine Foods' Shelf Life

The methods most used today to estimate the shelf life of foods are:

- Direct method: These are real-time studies that consist of storing the product under conditions similar to those that it will actually face, to monitor its evolution in regular intervals of time. The main advantage of this method is that it creates a very accurate estimation of the time it takes for a product to deteriorate; however, they are studies that usually take a long time and do not consider the fact that storage conditions of a product are not always stable over time.

- Challenge test: This method consists of experimentally introducing pathogens or microorganisms into the food during the production process, so that the product is exposed to the real conditions it will suffer in real life. The main disadvantage of this type of test is that the effects caused by the studied parameters are the only things analysed, and the fact that the product can be faced with multiple factors at the same time is not addressed. In addition, they are studies that are quite complex and difficult to implement.

- Predictive microbiology: This methodology studies the different microbial responses of foods to varying environmental conditions, based on mathematical and statistical models, in order to predict the behavior of the microorganisms in the product. This type of study, widely used when developing a new product, does consider the possible changing conditions of a product, however, its major limitation is that it implies greater complexity for the manufacturer and that the results correspond to a simulation, which may not be accurate.

- Accelerate shelf life tests: In these tests, conditions such as temperature, oxygen pressure or moisture content are modified to accelerate spoilage reactions of a food. These predictions allow one to predict the behavior of foods in certain conditions and to estimate how they will evolve under certain storage conditions. Accelerated tests allow the inclusion of changing environmental conditions and concentration variations of the ingredients that they are composed of. These studies are very versatile, low cost for the manufacturer and allow for the comparison of different scenarios. Obviously, since it is not an exact representation of reality, there is some margin of error in the obtained results.

- Survival method: It is a type of study that is based on the opinion of the consumer about the physical characteristics of the product. It consists in knowing the attitude of people towards the same product with different dates of manufacture, to determine if they would consume it or not. This method seeks to establish a relationship between the shelf life and the perceived quality of the product. Although it is not a method to accurately estimate the shelf life, it is important to do it in a complementary way to establish the best by date of a product.

Shelf Life Dating

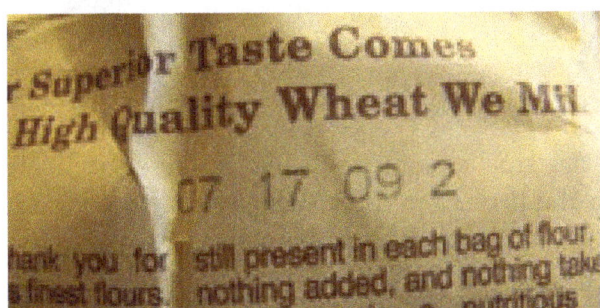

Because consumers feel they have the right to know about the product and its shelf life, various dating systems have been implemented in order to provide the consumer with information deemed necessary to purchase food items. The purpose of dating is to inform the consumer about the shelf life of the product.

Many different types of dating may appear on the label of various food products. Thus, shelf life dating can be evident in several ways.

- Pack date - the date on which the food was manufactured.

- Display dating - the date the food was placed on the store shelf.

- Sell by/pull date - the date the food must be sold or removed from the shelf.

- Best if used by dating - the date of the maximum quality of the product.

- Expiration date/use by date - the date the food must be consumed or discarded.

Food Deterioration

Foods are described as spoiled if organoleptic changes make them unacceptable to the consumer. These organoleptic characteristics may include changes in appearance (discoloration), the development of off-odours, slime formation, changes in taste or any other characteristic which makes the food undesirable for consumption. Whilst endogenous enzymatic activity within muscle tissue post-mortem can contribute to changes during storage, it is generally accepted that detectable organoleptic spoilage is a result of decomposition and the formation of metabolites caused by the growth of microorganisms.

The signs that food is spoiling are:

Odour: "off odours" are smells (sometimes like rotten eggs) that are produced when bacteria break down the protein in food, (usually fatty foods). This process is called putrefaction. Taints due to flavor change may also occur.

- Sliminess: Food becomes slimy as the bacterial population grows.

  Moulds may also form slimy whiskers.

- Discolouration: Foods can become discoloured by microbial growth.

  Some moulds have coloured spores that give the food a distinctive colour, for example, black pin mould on bread, or blue and green mould on citrus fruit and cheese.

- Souring: Foods go sour when certain bacteria produce acids. A common example is when milk sours from the production of lactic acid.

- Gas: Bacteria and yeasts often produce gaseous by-products that can affect food. It may be noticed that meat becoming spongy, or packages and cans swelling or having a popping or fizzing sound on opening.

According to the cause of the spoilage, types of food spoilage fall into two major categories. Microbial spoilage is caused by microorganisms and their products; non-microbial spoilage can be caused by foreign material in the foodstuff or by enzymes that occur in the foodstuff naturally.

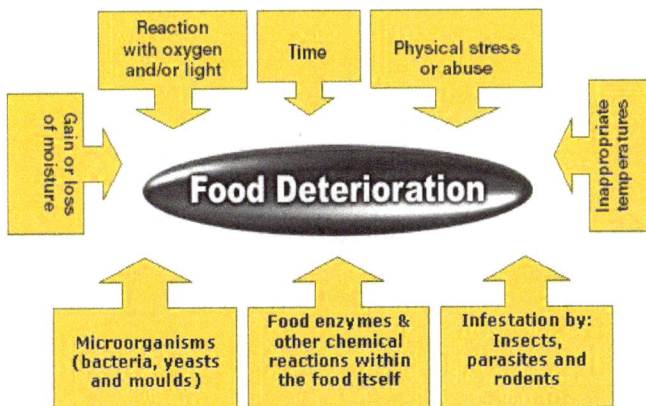

Microbial Spoilage

Spoilage of any particular food will be by those organisms most suited to the conditions in and around that food. The three main groups of concern are as mentioned below.

1. Bacteria:

Bacteria are the main and an important cause of food spoilage. They thrive where food and water are present and the temperature is suitable, as in the nose, throat, skin, bowel and lower urinary tract of man and animals. They are single cell organisms usually having a definite outer envelope or capsule for protection. They multiply by dividing into two, which can occur very quickly, (e.g. every 20 minutes). They can actively move and some link themselves together in chains or in bunches.

To resist harm, some bacteria can form spores (tough reproductive cells that are able to survive under adverse conditions), that can resist damage by heat (as in cooking), by cold (as in freezing) and by chemicals such as disinfectants. A spore can survive in dust, on vegetation and in soil for weeks, months or even years until it finds itself in a suitable environment for growth.

Figure below shows a typical growth curve. Four distinct phases occur in the growth curve: lag; log or growth phase; stationary phase; and death phase.

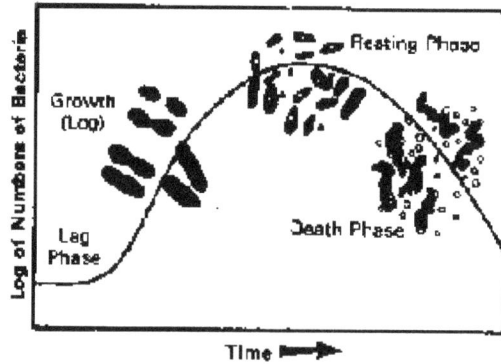

Figure: Growth curve

Bacteria need about four hours to adapt to a new environment before they begin rapid growth. In handling food, this means there is less than four hours to make a decision to cool the food, heat it, or eat it. For example, when chickens arrive at the dock of a fast food outlet, or at a restaurant or at home, it must be decide whether to heat and eat them, to refrigerate them at a low temperature (chickens freeze at 28 degrees F) for a short period of time, or whether to wrap and freeze the chicken for a longer period of time. If it is not decided, the bacteria will enter the log phase of growth where bacteria grow rapidly and cause food to spoil. Bacteria produce the slime, toxins, off colors and odors associated with food spoilage in the log phase of growth. It should be remembered that the four hours bacteria remain in the log phase is approximate and cumulative.

As microorganisms grow, they tend to form colonies. These colonies are made up of millions of individual cells. Once a colony forms, the food available to each cell is limited and excretions from these millions of cells become toxic to a microbe. This is the stationary phase. Some of the cells now begin to die.

2. Viruses:

Viruses are organisms much smaller than bacteria. In their pre-infective stage they are just like a chemical with none of the requirements for life, but once in a living cell they take over and begin to multiply. They can grow only in living tissue, but can be carried in food from one person to another.

3. Fungi:

Yeasts are single cell organisms much larger than bacteria and can be found in the soil, on plants and on the skin and body of man. They multiply by forming offspring as buds which grow and then detach themselves. Some can produce disease, some cause skin infections in man and others cause diseases in plants. Some yeasts spoil food, but beneficial uses are in the making of beer, wine and bread.

Moulds grow as single cell filaments that can branch together making a strongly knit structure like a mat, that can often be seen with the naked eye. Usually they look fluffy, being a familiar sight on foods like jam, cheese and bread. They multiply by producing clusters of dry spores which are blown by the air like seeds. Many moulds spoil food and a few can cause disease in plants and man, but beneficial uses are in the ripening of cheeses and production of antibiotics.

Control of Microbial Spoilage

Microbial spoilage is the major cause of food spoilage. It occurs as a result of contamination of food by microorganisms, provision of a suitable environment for their growth, and degradation of the foodstuffs. The micro-organism can grow:

- At temperatures between -7 to around 70°C.

- Over a pH range from 0 to 11.

- In the presence or absence of oxygen.

- At water activities above about 0.6.

To retard microbial buildup, the following parameters must be controlled:

- Source: Foodstuffs are naturally contaminated with microorganisms. To keep numbers of microorganisms as low as possible, fresh foods should be washed such as fruits and vegetables to physically remove as many microorganisms as possible. Processors of foods must keep their plants and equipment as clean as possible to provide clean work surfaces that come in contact with food. Every effort should be made to keep the initial numbers of microorganisms as low as possible.

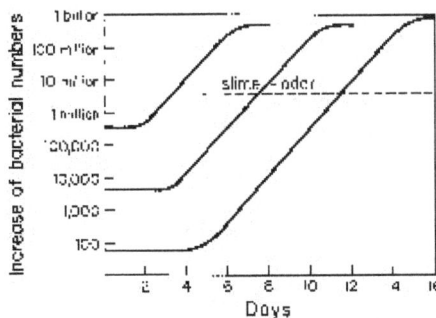

Figure: Variation in Bacterial number with time

Figure shows why it is needed to keep initial numbers low. A food product that starts with 100 microorganisms per gram may have a shelf life of 12 days before it develops off odors, slime and spoilage. When the initial number is 5,000 per gram, the shelf life of that same foodstuff may be shortened to seven days. Since so much depends on the initial number of bacteria, temperatures and handling practices, a specific shelf life for a category of food products is difficult to determine.

Initial numbers can be kept low by practicing good personal hygiene, by sanitizing equipment and controlling temperature, and by using chlorinated water where possible.

- Food: Like every other living thing, bacteria require food to live. They need only very small quantities. Some protein or fat left on the wall of a processing plant, grease on the blade of a knife or saw, or food residues on the wheel of a can opener or on a cutting board are a feast for microorganisms as well as for larger pests. Any equipment that may contact food (food contact surfaces) must be cleaned thoroughly.

- Moisture: Every living thing requires moisture, and bacteria are no exception. Food that requires refrigeration is usually very high in moisture content. Moist food left over for long periods of time provides adequate moisture for growth.

Figure below shows a typical bacterial cell. Its surface is rough, similar to a sponge's surface; its only means of obtaining food is by absorption similar to that of a sponge. Enzymes manufactured inside the cell move out onto foodstuffs, combine with the food, and return to the cell. This process cannot be accomplished without moisture. This is why foods such as dried milk, dried soups and cereals do not spoil microbiologically. The bacteria are present, but they can't eat.

Different bacteria require different temperatures for maximum growth. Some bacteria, called psychrophiles, will grow at refrigerated temperatures. Others, called mesophiles, will only grow at moderate temperatures. Warm-loving bacteria, called thermophiles, grow at temperatures above 140 degrees F. At temperatures above and below the optimum, they grow and reproduce at a slower rate. Food spoilage bacteria grow best at environmental temperatures of 70 to 100 degrees F. A generation time is the amount of time it takes for a bacterium to reproduce itself. The shorter the generation time, the faster food spoilage will occur.

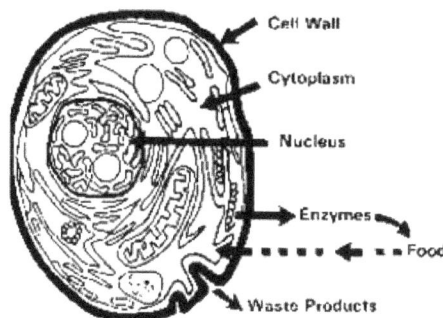

Figure: Typical bacterial cell

Non-microbial Food Spoilage

Food may spoil as a result of chemical changes within the food itself or by a reaction between the food and the packaging material. Rancidity is caused by a chemical reaction that breaks down the fatty acids in fat to smaller molecular weight fatty acids and, at the same time, releases certain odiferous products. Washers, bolts, nuts and various other items have been found in canned foods. This generally occurs when a maintenance person makes a repair on the line and uses a can for holding parts. The can stays on the line and is filled with the product. Although the product is retorted and sterile, it is aesthetically undesirable to find metallic parts in canned foods.

Enzymatic Spoilage

Enzymes are chemicals produced by all living things. They help speed up or slow down chemical reactions, act as transports for foods, and are a normal constituent of foods. For instance, as a banana matures, the color changes from green to yellow to brown to black. The change is caused by the enzymes (chemicals) in the banana. The ripening, then softening, of other fruits such as apples, peaches and tomatoes is another example of enzymatic action. Enzymes can be inactivated by heat, which is the reason for blanching vegetables; or they can be inactivated by cold temperatures below 40 degrees F, which is the reason for placing vegetables under refrigeration.

Bacteria also produce enzymes that break down food and allow them to obtain nutrients through their cell walls. Therefore, lowering the temperature reduces the rate of enzyme action as well as the rate at which bacteria can multiply. Refrigeration increases the time required to spoil food.

As the number of bacteria increases, the amount of enzymes produced increases. Higher temperatures can cause increased enzymatic activity. With large numbers of bacteria and high temperatures, a food will spoil very rapidly. When bacterial contamination is high and the storage temperature is low, a food will keep for a moderate period of time; when the bacterial contamination is low and the storage temperature high, food will keep for a moderate period of time. However, if the contamination of bacteria is low and the storage temperature kept low, the food product will have the longest possible shelf life.

Food Additives

Food additives are chemical substances added to foods to improve flavor, texture, color, appearance and consistency, or as preservatives during manufacturing or processing. Herbs, spices, hops, salt, yeast, water, air and protein hydrolysates are excluded from this definition.

Confusion has arisen regarding whether or not vitamins and minerals are considered to be food additives. In Australia, Food Standards of Australia and New Zealand (FSANZ) classifies vitamins and minerals as a distinct set of food additives, and they are regulated separately in their own standard. For example, the fortification of vitamin C in fruit juice or calcium in milk products is aimed at improving the nutritional quality of food products.

Food additives can be extracted from natural sources. Vitamin C, or ascorbic acid, (300) is extracted from fruit, and lecithin (322) from egg yolks. Food additives can also be synthesised in a laboratory. Synthetic compounds from the laboratory either do not occur in nature, or are chemically identical to natural materials (known as nature-identical).

In many countries, the use of food additives is regulated, and food additives must be declared on food labels by using their chemical names or numbers. In Australia, food additives are assessed by FSANZ before they are allowed to be used. Before FSANZ approves the use of any new additive in a particular food, they ensure that:

- The additive is safe to consume (at the requested level in that particular food);

- There are good technological reasons for the use of the additive; and

- Consumers will be clearly informed about its presence.

If no adverse health effects are demonstrated on the requested use, FSANZ will approve the food additive and recommend a maximum level of food additive permitted in particular foods. Generally, food additives approved by FSANZ are safe to consume without any adverse reactions, however, some people are sensitive to particular food additives in common use.

Coding of Food Additives

The food additive coding system was developed by the European Community (EC). The European food additive code numbers are prefixed by 'E' (e.g. E223). These E-numbers indicate the food additives that are approved for use in Europe.

Countries outside Europe use the numbers but do not add the E prefix. For example, acetic acid is written as E260 on food products sold in Europe, but it is known as additive 260 in Australia. Additive 103, alkanet, is approved for use in Australia and New Zealand, but is not approved for use in Europe and so does not have an E number.

Classification of Food Additives

Different types of food additives and their functions are listed in table:

Table: Classes of food additives

Class of additive	Function	Examples
Anti-caking agents	Keep powdered products (e.g. salt) flowing freely when poured	• Bentonite (558), • Calcium aluminium silicate (556), • Calcium silicate (552)
Anti-foaming agents	Reduce or prevent foaming in foods	• Polyethylene glycol 8000 (1521), • Triethyl citrate (1505)
Antioxidants	Retard or prevent the oxidative deterioration of foods	• Butylated hydroxyanisole (320), • Ascorbyl palmitate (304), • Calcium ascorbate (302)
Artificial sweeteners	Impart a sweet taste for fewer kilojoules/calories than sugar	• Sorbitol (420), • Alitame (956), • Aspartame (951), • Saccharin / calcium saccharin (954)
Bleaching agents	Whiten foods	• Chlorine (925), • Chlorine dioxide (926), • Benzoyl peroxide (928)
Bulking agents	Increasing the bulk of a food without affecting its nutritional value	• Ammonium chloride (510), • Isomalt (953), • Polydextrose (1200)
Colourings	Add or restore colour to foods	• Curcumin (110), • Brilliant blue FCF (133), • Tartrazine (102)
Colour retention agents	Retain or intensify the colour of a food	• Ferrous gluconate (579)

Emulsifiers	Prevent oil and water mixtures separating into layers	• Lecithin (322), • Sorbitan monostearate (491), • Ammonium salts of phosphatidic acids (442)
Enzymes	Break down foods (e.g. ferment milk into cheese)	• α-amylase (1100), • Lipases (1104), • Proteases (papain, bromelain, ficin) (1101)
Firming agents	Strengthen the structure of the food and prevent its collapse during processing	• Calcium chloride (509), • Calcium gluconate (578), • Calcium sulphate (516)
Flavour enhancers	Improve the flavour and/or aroma of a food	• Calcium glutamate (623), • Disodium 5′-ribonucleotides (635), • Ethyl maltol (637)
Food acids	Maintain a constant level of sourness in a food	• Acetic acid (260), • Citric acid (330), • Fumaric acid (297)
Flour treatment agents	Improve flour performance in bread making	• Sodium metabisulphite (223), • Ammonium chloride (510), • Potassium bromate (924)
Glazing agents	Impart a shiny appearance or provide a protective coating to a food	• Beeswax, white and yellow (901), • Carnauba wax (903), • Shellac (904)
Gelling agents	Thicken and stabilize various foods (e.g. jellies, deserts and candies)	• Agar (406), • Calcium alginate (404), • Carrageenan (407)
Humectants	Prevent foods from drying out (e.g. dried fruits)	• Glycerin or glycerol (422), • Lactitol (966), • Oxidised polyethylene (914)
Mineral salts	Improve the texture of a food (e.g. processed meats)	• Cupric sulphate (519)
Preservatives	Protect against deterioration caused by microorganisms	• Sodium nitrate (251), • Benzoic acid (210), • Sodium benzoate (211)

Propellants	Gases which help propel a food from a container	• Carbon dioxide (290), • Nitrogen (941), • Nitrous oxide (942)
Sequestrants	Bind and remove unwanted minerals that cause oxidation	• Potassium gluconate (577)
Stabilisers	Maintain the uniform dispersion of substances in a food	• Xanthum gum (415), • Guar gum (412), • Bleached starch (1403)
Thickeners	Improve texture and maintain uniform consistency	• Tannins (181), • Sodium alginate (401), • Pectins (440)
Vitamins	Restore vitamins lost in processing and storage	• B vitamins, including niacin • Vitamin C • Vitamin E

Looking for Additives in Foods

Food additives can be found in the ingredients list on the food label. Food labelling enables consumers to identify the presence of additives in packaged food and to make an informed choice about the foods they buy.

According to FSANZ, food additives are required to be identified by their class name, followed by an individual name or code number. To simplify the food labels, a code number is used to replace the individual name of the food additives. Food additives may be listed in food labels as, for example, thickener (guar gum) or thickener (412).

Differences between Natural and Artificial Coloring

Artificial or synthetic colors are synthesized in the laboratory. They can be chemically identical to colors that occurs naturally. Natural colors are derived from natural or bio-genic sources (e.g. animal, vegetable or mineral). Both natural and artificial colors are used in food products like ice creams, confectionery, biscuits, sweet meats, fruit drinks, seasonings, pharmaceutical tablets and syrups.

The use of color additives which might cause cancer or hyperactivity in children has raised concern among consumers. It is possible, but rare, to have an allergic-type reaction to a color additive. For example, tartrazine (102) is an artificial coloring that has been associated with allergic reactions in some rare instances. Reactions have ranged from rashes and swelling to asthma, and possibly even to behavioral changes. Individuals should read food labels and avoid certain color additives if they experience allergic reactions.

Safety Regarding Consumption of Food Additives

Although safety assessments of food additives are carried out by FSANZ before the food additives are approved for use, food additives can still induce adverse reactions in some sensitive individuals. According to FSANZ, it does recognize the adverse reactions to food additives in a small proportion of the population. These reactions are not the same as allergies, but may include rashes and swelling of the skin, irritable bowel symptoms, behavioral changes in children, and headaches.

Two major groups of food sensitivity are known as food allergy and food intolerance. Food allergies are abnormal immunologic responses to a particular food or food component. In contrast, food intolerances are non-immunologic responses. Generally, total avoidance of the culprit food is necessary for true food allergies. Food intolerances can be managed by limiting the amount of the food or food ingredient that is eaten. Total avoidance is usually not necessary for food intolerances.

Some commonly used food additives that tend to induce adverse reactions are mentioned below.

Aspartame

Aspartame (951) is an artificial sweetener that is used to replace sugars in foods and beverages. The long term effects of aspartame on health have been studied intensively, but results were inconclusive. It is noted that aspartame induces carcinogenic effects in a dose-related manner. Contradictory results were shown in studies which reported that aspartame consumption in foods and beverages does not raise the risk of brain or other cancers.

Although inconclusive results were shown in several studies, FSANZ and other international regulatory agencies concluded that aspartame is safe to consume. Aspartame is approved for general use in tabletop sweeteners, carbonated soft drinks, yoghurt and confectionery.

The acceptable daily intake (ADI) of aspartame is currently 50 mg/kg body weight in the United States, and 40 mg/kg body weight in Australia and the European Union for both children and adults.

Benzoate

Sodium benzoate (211) is used as a food coloring and preservative in foods. Children who consumed a mixture of food colorings and preservatives from soft drinks and confectionery at high levels were found to be more hyperactive than those who did not have the colorings and preservatives. Colorings and preservatives can be minimized in diets by including lots of fresh fruits and vegetables and eliminating processed foods.

Monosodium Glutamate (MSG)

Monosodium glutamate (621) is often added to food as a flavor enhancer but it can also occur naturally in food. While in the past MSG has been implicated as the causative agent of Chinese restaurant syndrome (CRS) and asthmatic attacks there is insufficient evidence to support this at the levels consumed in food.

Nitrates

Nitrates or nitrites are added as a preservative, antimicrobial agent or color fixative to processed foods such as meats and cheese. Nitrate also occurs naturally in water, vegetables and plants. The human body converts nitrate in food into nitrite. Nitrite has been implicated in a variety of long term health effects, including gastric cancer.

Sulphite

Sulphite sensitivity is a food intolerant reaction. Sulphites exist in several forms (e.g. sodium and potassium metabisulphite, sodium and potassium bisulphite, sodium sulphite, and sulfur dioxide). Sulphite has many functions, including as a antimicrobial agent. It inhibits enzymatic and nonenzymatic browning, whitens foods, and serves as a dough conditioner. Manifestations of sulphite sensitivity include anaphylaxis and asthma.

Tartrazine

Tartrazine (102) is an approved artificial food color. Tartrazine has been implicated in the aggravation of both asthma and chronic urticaria in some people. However, the association of tartrazine in the provocation of asthma and chronic urticaria is controversial. Some studies have shown a cause-and-effect relationship, whereas other studies have not. Both asthma and chronic urticaria are chronic illnesses with symptoms that tend to flare up at unpredictable times.

Sensitivity of Individuals to Food Additives

Food additives used in food products are approved by FSANZ. However, some individuals are sensitive to specific food additives. The degree of sensitivity varies from person to person.

Children

There has been a debate regarding the detrimental effect of artificial colors and preservatives on the behavior of children. It has been suggested that artificial food colors and other preservatives may produce overactive, impulsive and inattentive behaviors in children. However, the sensitivity of each individual to artificial colors is different. Some children have positive changes of behavior when artificial colors are eliminated from their diet, while others do not.

Children who have hyperactive or hyperkinetic behavior are to a larger extent diagnosed with attention deficit hyperactivity disorder (ADHD). Although ADHD is normally present at birth and tends to run in families, a study showed that food additives also predispose school-aged children to hyperactive behaviors. A child with ADHD has difficulty focusing his attention or engaging in quiet passive activities, and has educational difficulties, especially in relation to reading.

Advantages

Some additives improve or maintain the food's nutritive value. Vitamins a, c, d, e, thiamine, niacin, riboflavin, pyridoxine, folic acid, calcium carbonate, zinc oxide- and iron are often added to foods such as flour, bread, biscuits, breakfast cereals, pasta, margarine, milk, iodized salt and gelatin desserts. Instead of vitamin c, you may see ascorbic acid listed. Alpha-tocopherol is another name for vitamin e, and beta carotene is a source of vitamin a. In addition to providing nutrients, food additives can help reduce spoilage, improve the appearance of foods and increase the availability of a variety of foods throughout the year.

Disadvantages

Some food additives can potentially cause harmful side effects. For example, butylated hydroxyanisole, commonly known as bha, is a preservative used in foods including potato chips, crackers, beer, baked goods and cereal. It has been classified by the u. S. Department of health and human services as a preservative" reasonably anticipated to be a human carcinogen" sulfites, which are added to baked goods, wine, condiments and snack foods, could cause hives, nausea, diarrhea and shortness of breath in some people.

Colors as Additives

Coloring, in the form of dyes, pigments or other substances, is technically considered a food additive. These substances are often used to enhance color that's lost due to storage or processing. Pigments derived from natural sources, such as vegetables, minerals or animals, are exempt from certification. Man-made colors require testing by both the manufacturer and the fda to ensure they meet specific guidelines for purity.

Artificial preservatives can help your food last longer without becoming contaminated with food-borne illnesses, which is the reason they're found in so many different processed foods. Although all artificial preservatives used in the united states have been deemed" generally recognized as safe" by the U. S. Food and drug administration, not all of these additives are 100-percent safe for everyone. Some preservatives are associated with adverse effects, which can involve an unpleasant reaction in people sensitive to a particular additive or a potential increased risk for cancer.

Preserving Food

Artificial preservatives may act as antioxidants, make food more acidic, reduce the moisture level of food, slow down the ripening process and prevent the growth of microorganisms, all of which help the food last longer. This means you can make fewer trips to the store and have less food waste because the preservatives help minimize the amount of food you buy that goes bad before you can eat it.

Limiting Foodborne Illnesses

Approximately one out of every six americans get a foodborne illness each year, according to the centers for disease control and prevention. Without artificial preservatives to limit the spread of the organisms that cause these illnesses, this number might be even higher. Some of these illnesses, such as botulism, can be deadly.

Potential Adverse Reactions

Certain preservatives, including sulfites and sodium benzoate, may cause adverse reactions in a small percentage of the population. Sulfites help limit the growth of bacteria in wine and the discoloration of dried fruit, but can cause potentially deadly allergic reactions in sensitive individuals, including rashes, low blood pressure, diarrhea, flushing, abdominal pain, asthmatic reactions and anaphylactic shock. Sodium benzoate, also called benzoic acid, is used in acidic foods to keep microorganisms from growing. In sensitive individuals, it can cause asthma, hives.

Increased Cancer Risk

Although sodium benzoate is usually considered safe for people who aren't sensitive to it, when combined with ascorbic acid in acidic foods it can produce benzene, which may slightly increase your risk for leukemia and other types of cancer, according to the center for science in the public interest. Nitrates and nitrites, which are often used to preserve cured meats, such as lunch meat and hot dogs, may also increase your risk for certain types of cancers.

Some Food Additives and their Side-effects

Tartrazine (E102), which is primarily used by the soft drink industry, is one of the colours most frequently implicated in food intolerance studies. Adverse reactions to tartrazine seem to occur most commonly in subjects who are also sensitive to acetylsalicylic acid (ASA), a finding which was also observed by Feingold and his team. Depending on the test protocol followed, it has been found that between 10-40% of aspirin-sensitive patients are indeed usually also affected by tartrazine, the reactions including asthma, urticaria, rhinitis and, as previously mentioned, childhood hyperactivity.

This may come about because the chemical structure of the tartrazine molecule has similar features to those of benzoates, other azo compounds, pyrazole compounds and the hydroxy-aromatic acids, which also include salicylates. Furthermore, it has been established that the azo compounds can be reduced in the intestine and in the liver, indicating that one of the several routes through which these molecules, too small to be antigenic in themselves, may act as a hapten, thus conjugating a larger molecule to form an antigenic compound.

A major breakthrough in the understanding of the mechanisms involved in ASA intolerance came also with discovery that aspirin, including other non-steroidal anti-inflammatory drugs, inhibit the synthesis of prostaglandins, by selectively blocking the cyclo-oxinase pathway, resulting in an enhanced production of leucotrienes. An excessive leucotriene production in turn leads to vascular permeability, causing oedema and inflammation, which is directly associated with various airway constriction disorders, including asthma.

One study found that an oral administration of 50mg tartrazine to 122 patients suffering from allergy-related disorders, evoked the following reactions; feeling of suffocation, weakness, heat sensation, palpitations, blurred vision, rhinorrhoea, pruritus and urticaria. Even though 50mg could be considered as a substantial dose, such a quantity of tartrazine could easily be consumed by an individual drinking only a few bottles of soft drinks per day.

Another carefully conducted double-blind placebo- controlled trial on 76 children diagnosed as hyperactive, showed that tartrazine and benzoates provoked abnormal behaviour patterns in 79% of them. In addition, a double-blind placebo-controlled trial on 10 hyperactive children when compared to controls, found that tartrazine increases urinary zinc secretion, and decreases serum and salivary zinc concentration in the hyperactives, with a corresponding deterioration in their behaviour. This phenomena was not found among the controls.

It was suggested therefore that tartrazine seems to act as a zinc chelating agent in susceptible individuals. Furthermore, that zinc depletion may also be one of the potential causes of childhood hyperactivity.

Although tartrazine seems to be most frequently associated with adverse reactions, there are also other coloring agents which are known to cause mental and/or physical ill-effects.

Curcumin (E100), used mainly in flour confectionery and margarine, has been found to cause mutations in bacteria and when fed to pigs, it increased the weight of their thyroid glands causing, in high doses, severe thyroid damage.

Sunset Yellow (E110), used in biscuits, has been found to damage kidneys and adrenals when fed to laboratory rats. It has also been found to be carcinogenic when fed to animals.

Carmoisine (El22), used mainly in jams and preserves, was found by the US Certified Color Manufacturers Association to be unavoidably contaminated with low levels of beta- napthylamine, which is a well known carcinogen; it has also been found to be mutagenic in animal studies.

Amaranth (El23) has been found, when fed to laboratory rats, to cause cancer, birth defects, still births, sterility and early foetal deaths. Subsequent work has also found that amaranth can cause female rodents to reabsorb some of their own foetuses.

Ponceau 4R (El24), used mainly in dessert mixes, has been found to exhibit a weak carcinogenic action.

Erythrosine (E127), used in candied cherries and childrens' sweets, has been found to act as a potent neurocompetitive dopamine inhibitor of dopamine uptake by nerve endings when exposed in vitro on a rat brain. Other studies have shown that erythrosine can have an inhibitory action also on other neurotransmitters, resulting in an increased concentration of neurotransmitters near the receptors, thus functionally augmenting the synaptic neurotransmission.

There is now some evidence that a reduced dopamine turnover may lead to childhood hyperactivity. Similar findings have been linked with a reduction of noradrenaline. Erythrosine also has been found to have a possible carcinogenic action when tested on animals.

Caramels (E150), of which over 100 different formulations are currently in use, are widely used by the cola drinks industry, as well as the beer and alcohol industry. It is also used as a coloring agent in crisps, bread, sauces, gravy browning etc.

The main recurring problem about the safety of caramels concerns the presence of an impurity called 4-Methylimidazole, produced by processes using ammonia, which leads to convulsions when fed to rats, mice and chicks. It has been also found that ammoniated caramels can affect adversely the levels of white blood cells and lymphocites in laboratory animals. Furthermore, a study on rabbits provided evidence that even small doses of ammoniated caramels seem to inhibit the absorption of vitamin B_6.

Brown FK is mainly used as a colouring agent in fish, such as kippers. Two of the primary metabolites of this colouring have been found to act as a cardiotoxin. It has been also observed, when fed in the long term to mice, to cause potentially hazardous nodules to form in the liver. Furthermore it has been found to cause mutations in some bacteria, implying that it may also act as a mutagenic and/or carcinogenic agent in humans.

Preservatives/Antioxidants

Benzoates (E210-E219), used mainly in marinated fish, fruit- based fillings, jam, salad cream, soft drinks and beer, have been found to provoke urticaria, angioedema and asthma. Furthermore, they have also been directly linked with childhood hyperactivity.

Sulphites (E220-E227), used mainly in dried fruits, fruit juices and syrups, fruit-based dairy deserts, biscuit doughs, cider, beer and wine, have been linked with pruritus, urticaria, angioedema and asthma. When fed to animals, sulphites have also been found to have a mutagenic action.

Nitrates and nitrites (E249-E252), used in bacon, ham, cured meats, corned beef and some cheeses, have been found to cause headaches in susceptible individuals. In addition, these chemicals have been linked with cancer both in animal and human studies. They have also been found to be mutagenic when fed to mammals.

Butylated hydroxyanisole - BHA (E320), used in soup mixes and cheese spread, has been found to be tumour-producing when fed to rats. In human studies it has been linked with urticaria, angioedema and asthma.

Monosodium glutamate (MSG), a flavor enhancer, used in savory foods, snacks, soups, sauces and meat products, has been associated with a conjunction of symptoms in susceptible individuals, such as severe chest and/or facial pressure and overall burning sensations, not unlike a feeling that the victim is experiencing a heart attack.

MSG has been also found to precipitate a severe headache and/or asthma in susceptible individuals. In susceptible children MSG has been linked with epilepsy-type "shudder" attacks. In animal studies it has been found to damage the brains of young rodents.

Sweeteners

Saccharin, used as sweetening tablets and widely used by the soft drink and sweet food industry, has been shown to produce cancer when tested on animals. Saccharin has also been found to be mutagenic and growth inhibiting, as well causing congenital malformations in animal studies.

The fact that any substance which has been found to be carcinogenic also seems to have a mutagenic action, was established by testing 300 different carcinogenic chemicals for mutagenicity. The results showed that of the 300 carcinogenic chemicals tested, 90% were also found to have a mutagenic action.

Aspartame, of which the key ingredient is the amino acid phenylalanine, is also widely used by the soft drink and sweet food industry. When fed to rats, aspartame was found to double the level of phenylalanine in their brains, which re-doubled when other carbohydrates were consumed at the same time. This combination was found to give a great rise in brain tyrosine, followed by a considerable reduction in brain tryptophan levels. Low tryptophan levels have been directly linked with both aggressive and violent behaviour.

Furthermore, as dietary tryptophan acts as a precursor for serotonin (5-hydroxy-trytamine, 5HT), reduced tryptophan levels will also result in a reduction of brain serotonin levels, which has been directly linked with both hyperactive and aggressive behaviour.

Sucrose/table sugar, which is a simple molecular substance artificially refined from complex carbohydrates, thus called a refined carbohydrate, can be found in most of our foods. An excessive refined carbohydrate consumption has been directly associated with a high incidence of both criminal and antisocial behaviour.

Schoenthaler and his team conducted several double- blind, design-over studies among thousands of incarcerated juvenile offenders, finding a clear correlation between high sucrose/junk food intake and the incidence of antisocial behaviour. In all studies the primary dietary revision to reduce sugar consumption was organised by simply replacing sugary drinks and junk food snacks with fruit juices and nutritious snacks, such as nuts and fresh fruits.

After implementing this simple dietary policy with 276 incarcerated offenders, informal disciplinary actions were lowered 48%, when contrasting the twelve months before, and after nutritional revision. Assault and battery was lowered 82%, theft 77%, horseplay 65% and refusal-to-obey-an-order 55%.

When similar diet policy was designed for 1382 offenders confined in three different juvenile institutions, there was a clear 25% reduction in rule violations. All 1382 juveniles served as their own controls and the length of the pre- and post-intervention period lasted for three months each.

Similar findings were observed when the behavior of 2005 incarcerated offenders was analyzed for 24 months. In the second half of the experiment i.e. after 12 months of the initial observation period, the offenders were no longer allowed foods/drinks containing sucrose or artificial additives; instead they were offered nuts, fruit, and fruit juices. After implementing the low-sugar diet policy the incidence of rule violations fell 21%, assaults and fights 25% and general disruptions 42%.

The consumption of sucrose/additive-rich foods was not only seen to worsen the behavior of young offenders, but when given to children diagnosed as hyperactive, these foods seemed greatly to increase their restless and destructive behavior.

Similar results were established among a group of normal pre-school children, as sucrose was found significantly to correlate with their "inappropriate behavior" pattern.

An excessive refined carbohydrate consumption can also lead in susceptible individuals to a disordered carbohydrate metabolism, especially to reactive hypogly-caemia, which in turn has been found to be particularly prevalent among violent offenders. Reactive hypoglycemia has also been associated with diverse personality and psychiatric disorders, such as neuroses, panic attacks, agoraphobia and schizophrenic episodes.

WHO Response

Evaluating the Health Risk of Food Additives

WHO, in cooperation with the Food and Agriculture Organization of the United Nations (FAO), is responsible for assessing the risks to human health from food additives. Risk assessment of food additives are conducted by an independent, international expert scientific group – the Joint FAO/WHO Expert Committee on Food Additives (JECFA).

Only food additives that have undergone a JECFA safety assessment, and are found not to present an appreciable health risk to consumers, can be used. This applies whether food additives come from a natural source or they are synthetic. National authorities, either based on the JECFA assessment or a national assessment, can then authorize the use of food additives at specified levels for specific foods.

JECFA evaluations are based on scientific reviews of all available biochemical, toxicological, and other relevant data on a given additive – mandatory tests in animals, research studies and observations in humans are considered. The toxicological tests required by JECFA include acute, short-term, and long-term studies that determine how the food additive is absorbed, distributed, and excreted, and possible harmful effects of the additive or its by-products at certain exposure levels.

The starting point for determining whether a food additive can be used without having harmful effects is to establish the acceptable daily intake (ADI). The ADI is an estimate of the amount of an additive in food or drinking water that can be safely consumed daily over a lifetime without adverse health effects.

International Standards for the safe use of Food Additives

The safety assessments completed by JECFA are used by the joint intergovernmental food standard-setting body of FAO and WHO, the Codex Alimentarius Commission, to establish levels for maximum use of additives in food and drinks. Codex standards are the reference for national standards for consumer protection, and for the international trade in food, so that consumers everywhere can be confident that the food they eat meets the agreed standards for safety and quality, no matter where it was produced.

Once a food additive has been found to be safe for use by JECFA and maximum use levels have been established in the Codex General Standard for Food Additives, national food regulations need to be implemented permitting the actual use of a food additive.

References

- Most-commonly-used-methods-determining-shelf-life-food: natural.btsa-es.com, Retrieved 31 March 2018

- Food-spoilage-and-control: agropedia.iitk.ac.in, Retrieved 23 May 2018

- Food-additives, lifestyles: myvmc.com, Retrieved 13 July 2018

- Advantages-and-disadvantages-of-food-additives: lybrate.com, Retrieved 28 June 2018

- Fact-sheets, detail, food-additives: who.int, Retrieved 18 July 2018

Food Packaging Materials

The material used for food packaging directly impacts its durability for consumption. This chapter closely examines the different materials used for food packaging, such as metal packaging materials, glass packaging materials, plastic packaging materials, and edible and bio based packaging materials, for a comprehensive understanding of food packaging.

Packaging Materials

The first packages served as containers, and their principal function was to hold food and water. They were probably taken directly from nature, such as leaves and shells. Later, containers were fashioned from natural materials: wooden logs, woven plant fibers, pouches made from animal skins. The next containers developed by early societies were clay pots, which date back to 6000 B.C. The first known pottery is from Syria, Mesopotamia, and Egypt. Besides being functional, clay bowls, vases, and other vessels were an artistic medium that today provide important clues regarding the culture and values of ancient peoples. Although no longer a significant packaging medium, clay still continues to have a major artistic value.

Paper based Packaging Products

The use of paper and paperboards for food packaging dates back to the 17th century with accelerated usage in the later part of the 19th century (Kirwan 2003). Paper and paperboard are sheet materials made from an interlaced network of cellulose fibers derived from wood by using sulfate and sulfite. The fibers are then pulped and/

or bleached and treated with chemicals such as slimicides and strengthening agents to produce the paper product. FDA regulates the additives used in paper and paperboard food packaging (21 CFR Part 176). Paper and paperboards are commonly used in corrugated boxes, milk cartons, folding cartons, bags and sacks, and wrapping paper. Tissue paper, paper plates, and cups are other examples of paper and paperboard products.

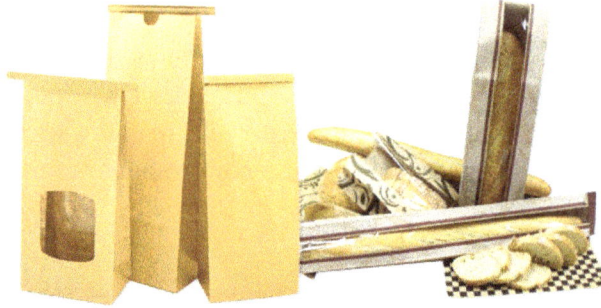

Paper

Plain paper is not used to protect foods for long periods of time because it has poor barrier properties and is not heat sealable. When used as primary packaging (that is, in contact with food), paper is almost always treated, coated, laminated, or impregnated with materials such as waxes, resins, or lacquers to improve functional and protective properties. The many different types of paper used in food packaging are as follows:

- Kraft paper—Produced by a sulfate treatment process, kraft paper is available in several forms: natural brown, unbleached, heavy duty, and bleached white. The natural kraft is the strongest of all paper and is commonly used for bags and wrapping. It is also used to package flour, sugar, and dried fruits and vegetables.

- Sulfite paper—Lighter and weaker than kraft paper, sulfite paper is glazed to improve its appearance and to increase its wet strength and oil resistance. It can be coated for higher print quality and is also used in laminates with plastic or foil. It is used to make small bags or wrappers for packaging biscuits and confectionary.

- Greaseproof paper—Greaseproof paper is made through a process known as beating, in which the cellulose fibers undergo a longer than normal hydration period that causes the fibers to break up and become gelatinous. These fine fibers then pack densely to provide a surface that is resistant to oils but not wet agents. Greaseproof paper is used to wrap snack foods, cookies, candy bars, and other oily foods, a use that is being replaced by plastic films.

- Glassine—Glassine is greaseproof paper taken to an extreme (further hydration) to produce a very dense sheet with a highly smooth and glossy finish. It is used as a liner for biscuits, cooking fats, fast foods, and baked goods.

- Parchment paper—Parchment paper is made from acid-treated pulp (passed through a sulfuric acid bath). The acid modifies the cellulose to make it smoother and impervious to water and oil, which adds some wet strength. It does not provide a good barrier to air and moisture, is not heat sealable, and is used to package fats such as butter and lard.

Paperboard

Paperboard is thicker than paper with a higher weight per unit area and often made in multiple layers. It is commonly used to make containers for shipping—such as boxes, cartons, and trays—and seldom used for direct food contact. The various types of paperboard are as follows:

- White board—Made from several thin layers of bleached chemical pulp, white board is typically used as the inner layer of a carton. White board may be coated with wax or laminated with polyethylene for heat sealability, and it is the only form of paperboard recommended for direct food contact.

- Solid board—Possessing strength and durability, solid board has multiple layers of bleached sulfate board. When laminated with polyethylene, it is used to create liquid cartons (known as milk board). Solid board is also used to package fruit juices and soft drinks.

- Chipboard—Chipboard is made from recycled paper and often contains blemishes and impurities from the original paper, which makes it unsuitable for direct contact with food, printing, and folding. It is often lined with white board to improve both appearance and strength. The least expensive form of paperboard, chipboard is used to make the outer layers of cartons for foods such as tea and cereals.

- Fiberboard—Fiberboard can be solid or corrugated. The solid type has an inner white board layer and outer kraft layer and provides good protection against impact and compression. When laminated with plastics or aluminum, solid fiberboard can improve barrier properties and is used to package dry products such as coffee and milk powder. The corrugated type, also known as corrugated board, is made with 2 layers of kraft paper with a central corrugating (or fluting) material. Fiberboard's resistance to impact abrasion and crushing damage makes it widely used for shipping bulk food and case packing of retail food products.

Paper Laminates

Paper laminates are coated or uncoated papers based on kraft and sulfite pulp. They can be laminated with plastic or aluminum to improve various properties. For example, paper can be laminated with polyethylene to make it heat sealable and to improve gas and moisture barrier properties. However, lamination substantially increases the cost of paper. Laminated paper is used to package dried products such as soups, herbs, and spices.

Carton

Corrugated boxes simply refer to what is commonly known as: Carton. Corrugated boxes are the ones many probably consider as 'Carton' as it produces the large shipping, shoe & storage boxes. What a lot of people do not realize is that corrugated boxes also come in different types depending on the durability and strength of the box. Identifying a certain corrugated material, however, is easy. How do you determine the material? Through its corrugated medium (also known as fluting). Identifying a corrugated material is easy. It consists of 3 layers of paper, an outside liner, an inside liner and a corrugated medium (also known as fluting). The corrugated medium that gives it strength and rigidity.

The main raw material that is used to construct the corrugated board is most recycled paper, made on large high-precision machinery known as corrugators. These type of boards can reused and recycled again and again as a source of pulp fiber. Corrugated boards are of different types, single faced, double faced (single wall), twin wall, and triple wall. They can be used to make packaging with different characteristics, performances, and strength. The board is cut and folded into different sizes and shapes to become corrugated packaging. Other applications of corrugated board packaging include retail packaging, pizza delivery boxes, small consumer goods packages, and so forth.

Chipboard Packaging

Chipboard packaging is used in industries such as electronic, medical, food, cosmetic, and beverage. A chipboard basically is a type of paperboard that is made out of reclaimed paper stock. It can be easily cut, folded, and formed. It is a cost-effective packing option for the products.

It comes in various densities and strength is determined by how high the density of the material is. If one want images to be directly printed onto the chipboard, one can treat the chipboard with bleach sulfate, and with CCNB (Clay Coated News Back) which makes the material even more durable.

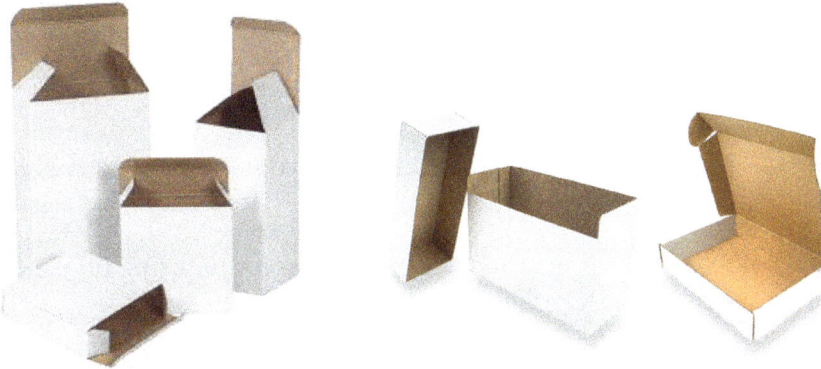

One cannot use chipboard packaging if the business deals with heavy items, since the chipboard sheet is a lightweight material, made for many grocery items such as cereal, crackers, tissue boxes, and so forth. Also, if the storage environment is heavy with moisture, chipboard can easily weaken, and result in discoloration and expansion.

Rigid Boxes

A rigid box is made out of highly condensed paperboard that is 4 times thicker than the paperboard used in the construction of a standard folding carton. The easiest real-world example of rigid boxes are the boxes that hold Apple's iPhones and iPads, which are 2 piece setup rigid boxes.

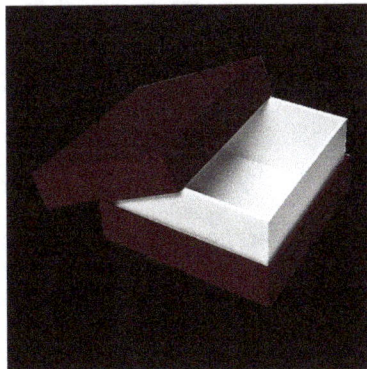

Compared to paperboard and corrugated boxes, rigid boxes are definitely among the most expensive box styles. The rigid boxes usually do not require dies that are expensive or massive machinery and are often hand-made. Their non-collapsible nature also gives them a higher volume during shipping, which easily incurs higher shipping fees.

These boxes are commonly used in merchandising cosmetics, jewelry, technology, and high-end luxury couture. It is easy to incorporate features such as platforms, windows, lids, hinges, compartments, domes, and embossing in a rigid box.

Liquid Packaging Board

Liquid tight packaging is designed with a special folding technique to ensure it is absolutely leak-proof and uses a recyclable coating system (RCS), which offers the highest barrier performance.

The design, in combination with the coated material, acts as a safeguards to ensure that no liquid will escape during transport and that your product arrives in optimal condition.

100% recyclable, liquid packaging is a single material solution using no plastics and is a reliable alternative to Bag-in-Box solutions.

Liquid packaging is food safe and suitable for liquid products such as cream concentrates or fruit pulp.

Features

- Bespoke to your exact size and performance requirements.

- Available in a wide range of corrugated cardboard and solid board grades.

- Uses a recyclable coating system for optimum liquid protection.

- Suitable for food contact.

- Leak-proof designs using a special folding technique.

- High quality print available to communicate your brand messages or product details.

Benefits

- Reduction in transit damage due to strength and stability of packaging.

- Environmentally friendly - Made from a renewable resource.

- Easy disposal - 100% recyclable in regular cardboard waste stream.

- Reduction in storage space and logistic costs as products can be delivered flat.

- Easy to use as no machines required to erect.

Sustainable Barrier Coatings

Liquid paperboard is providing a key market application for barrier coating materials that has a reduced environmental impact.

Several biopolymers have already been commercially trialled in coatings for paperboard. The most promising polymers identified were polyhydroxybutyrate (PHB) and polylactic acid (PLA). These two materials offer the best results, in terms of a consistent, thick and continuous layer, for extrusion coating, while other materials are better suited to dispersion and solvent casting.

Smaller Pack Sizes

Social changes are affecting packaging demand in unexpected ways. There is an increasing need for greater convenience and a growing demand for ready-to-drink and on-the-go foods and beverages. The result is a growing demand for smaller pack sizes in numerous food and beverage sectors.

Smaller packs consume more liquid paperboard to pack the same given quantity of product, leading to an increase in raw material consumption. Furthermore, these smaller packs consist of a greater surface area of material that requires more inks, adhesives and other consumables, stimulating value-added for converters.

Competing Materials

Competition from glass packaging and fluctuations in the price of aluminium are having a degree of influence on both liquid packaging board and multi-substrate flexible pouches.

Liquid packaging board faces competition from specific packaging formats and the latest of these is the clear retortable plastic can. This is likely to have a limited impact on the liquid paperboard market, but it does pose a threat to the retort carton, which has shown significant opportunities.

Recyclibility

Governments and beverage brand chains are increasingly pushing for recyclable hot beverage cups. A recent innovation in mid-2016 has seen Starbucks trial a fully recyclable cup as part of its efforts to reduce waste sent to landfill, as well as to improve its environmental profile.

Sustainability

A broad trend across the packaging industry is to minimize the environmental impact of packaging, which is stimulating both technical innovations and new business models across the value chain.

This interest includes consumers – especially millennials – brand owners, converters and regulators and spans a number of different means to realize the goal of furthering the use of sustainable, recyclable and compostable packaging into the mainstream.

Reduction of packaging waste can be achieved by various means, including lightweighting and by utilizing materials that are themselves sustainable, as in renewable and recyclable. This has been perceived to be an advantage of liquid paperboard, but new understandings of how polymer coatings inhibit this process, and the arrival of competing formats optimized for sustainability is threatening this.

An example of this competition is Coca-Cola's Plant Bottle concept. This is produced using up to 30% plant-based materials (biopolymers) including stems, barks and fruit skins, and Coca-Cola aims to produce all plastic bottles solely from recycled and bio-derived feedstocks , by 2020.

Healthy Eating Trends

In efforts to reduce welfare costs, various governments are endorsing healthier eating lifestyles and promoting healthier habits, which are being adopted by many consumers. The EU has targeted a 16% reduction in salt for processed foods, while the UK reduced the average consumption of salt by around 5% across 2005–2011.

Consumers have become more sensitive to prices, which has become evident in the strongly competitive food retail industry. In Western Europe, private labels are increasingly popular and are becoming a brand in their own right. The initial advantage that traditional brands had over private labels was the perceived quality differential, which still exists to differing extents in some countries, but is most evident in Italy. However, demand is changing, largely due to the poor economic climate. In addition, the improved quality branding of private labels has helped to put them on par with their branded equivalents.

Molded Pulp

Molded pulp is a type of product made from recycled paper that offers the benefits of cardboard or foam packaging without the drawbacks of non-sustainable materials. Sustainable packaging offers many benefits over traditional packing materials. Pulp products are created entirely from recycled paper, making it one of the most sustainable and environmentally friendly packaging solutions. This means that fewer trees are destroyed in creating more paper, and the paper that is used does not go into landfills as trash.

Manufacturing process consists of several steps:

- Recycled paper is collected from various sources and stored onsite until it is made into pulp moulded packaging.

- The recycled paper is sent through a pulping system, turning it into fiber pulp, the basic ingredient in our molded pulp tray products.

- Recycled water is added from a storage tank, turning the pulp into a slurry of paper and water.

- The pulp and water are dried through a vacuum system to extract excess water. The excess water is then pumped back to the recycled water tank. In this way, we reuse our water constantly and reduce the need for more water.

- The pulp material for the molded tray is formed and shaped into a unique configuration and dried completely to hold its shape. Our oven drying process uses heat exchangers. These reduce energy-use and makes our process more efficient.

Once the molded pulp has been processed and turned into convenient storage trays, it is shipped to customers for use as inner support shipping packaging. This molded pulp container can then be reused or recycled, continuing the cycle of sustainability.

Molds decide the shapes of final products, which are featured with elegant appearance, good stiffness, light weight and are easy to be stored and transported. Most important, raw material of these pulp molded products have a wide, cheap resource, and is easily available. Reed, straw, bamboo and various plant fibers can be utilized effectively. After application, all paper tableware can be 100% recycled because of their good biodegradability.

Due to excellent plasticity, any shape can be molded by a pulp moulding machine, all you have to do is just change dies.

Fast Food Tableware

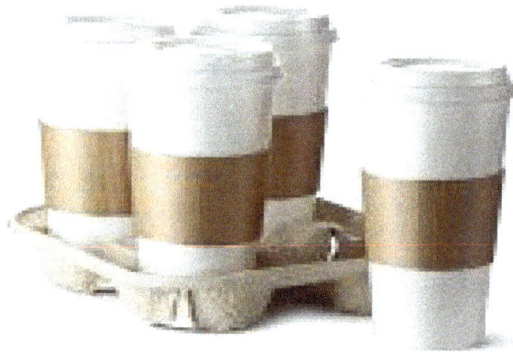

Pulp molded tableware can be applied to dishware, bowl, fast food tray, cup carrier and more. These molded fiber products are featured with elegant appearance, good stiffness, light weight and are easy to be stored and transported.

They are water & oil proof, and can be froze by refrigerator or heated up by micro-wave oven, which is suitable for eating habits of modern people and fast food processing. According to a study, one ton of paper pulp can produce 40 thousands of 600ml standard dishware.

Egg Tray/Carton

Due to porous material, unique egg-shaped curved structure, better ventilation, freshness preservation and good cushioning, molded pulp egg tray/carton is especially suitable for packaging of high volume egg transportation, no matter chicken, duck, goose or quail egg.

Statistic shows that egg broken rate reclined to 2% from 8%~10% by using egg tray instead of traditional packaging method. Logo or other characters can be printed on egg tray/carton for exhibit impression. Just choose the right type egg try machine for your egg tray business.

Fruit Tray

Paper pulp can be molded into the shape of fruit, and used as fruit tray.

Molded Pulp Packaging of peach, pear, apple, orange, pineapple, tomato, and fruit for export, can avoid damage due to collision, distribute respiratory heat of fruit, absorb evaporated water, suppress ethylene density, prevent fruit from going bad and prolong freshness. All these actions are irreplaceable than other packaging material.

Egg Carton

Egg Cartons are the cartons used for carrying eggs.

Its functions are:

- Protection function

  Egg trays can protect the egg from the external shock. It can also protect the eggs from damaging and going bad due to illumination and dampness so that the loss can be reduced.

- Convenient transportation

  The purpose of producing eggs is to let them get into market, which is a process of product circulation. In this process, eggs will undergo countless times of transportation, handling, storage, which requires the egg packing must be able to adapt to the process. Egg trays can satisfy this purpose.

- Increasing sale

  In today's market, great changes have taken place in the way of people's purchasing. With the improvement of people's material standard, more and more people begin to pursue a higher quality of life. The packaging quality of the products has a direct impact on sales. The egg is more typical. If we make the egg packaging exquisite and unique, it will increase the sales of eggs.

Metal Packaging Materials

Metal packaging is found everywhere in modern daily life due to its versatility and universal consumer approval. Its sustainability credentials combined with the outstanding characteristics listed below make it an ideal packaging solution that meets the changing demands of our society in the 21st century:

The advantages of the metal packaging:

- Keeps the content safe for longer.

- Prevents product waste.

- Keeps the packed nutrients clean from impurities.

- Extends the shelf life of the packed products.

- Ensures the protection form light, oxygen and bacteria – the perfect solution for modern lifestyle.

- The products remain fresh until the opening of the can.

Metal packaging with its extended shelf life provides protection against light, oxygen and bacteria and is an excellent solution for modern lifestyles. The can ensures products stay fresh until consumed contributing to value for money for consumers.

Metal continues to have immense potential for the development of innovative packaging solutions. The applications are unlimited, ranging from packs for chewing gum, microwaveable ready meals to aerosols and do-it-yourself (DIY) products.

Metal can solutions are continually being improved and developed to meet customer and consumer needs.

Hermetically closed, the metal packaging contributes to the complete preservation of the food from the moment of its closing.

Metal packaging allow the transportation and storage of harmful to people and environment materials.

Today, metal is used for a vast range of products and packaging solutions, yet designers continue to seek new opportunities to delight consumers. Cans come in many shapes, sizes and decorations for:

- Food

- Beverages

- Lifestyle and luxury products

- Personal hygiene

- Household products

- DIY

- Industrial

Convenience

Convenience remains a key driver in packaging development for consumer packaged goods. Changes in family structures, flexible working and commuting hours, resulted in changing of the consumption habits. Smaller household units, more working women and mothers and busier lifestyles, the packaging industry has been constantly evolving to satisfy these changing needs.

Metal packaging enhances product safety by being both tamper proof as well as virtually unbreakable. Its barrier properties, which keep the goodness in and impurities out ensure that the content safely reaches the consumer, free from contaminants.

Metal packaging satisfies new demands from consumers via easy opening, re-sealable packaging and through manageable portions which allow consumers to open only what they need. Designs for metal cans which can be used in microwave ovens extend the convenience of metal packaging for ready meals.

Canned foods are safe to eat for many years – ideal for today's meal or tomorrow's emergency. Cans do not need refrigeration or freezing. They provide total protection against light, oxygen and, being hermetically sealed, also protect food to keep the nutrients locked inside. Products are also protected against moister, dust, rodents and other contaminants.

Logistics

Cans are typically 35% lighter than 20 years ago, but they are still unbreakable.

High-speed filling, at up to 1,500 cans a minute, saves significant amounts of energy compared with other packaging solutions.

Most can-making facilities are close to or next to filling lines, ensuring efficient and flexible delivery.

Stackable and tapered cans cater for efficient long-distance supply. Empty cans can be stacked over eight meters high, making the best use of warehouse space.

Embossing

Embossing and de-bossing technologies enable can makers to produce innovate can designs.

Embossing is a technology which produces decorations or reliefs with a raised profile (from inside out) while de-bossing produces decorations with a depressed profile (outside-in).

The advantage of de-bossing is that existing external dimensions are maintained and therefore there is no need to change transportation and pallet space.

The former metal is registered with printed graphics for maximum visual and tactile impact. Images and graphics, logos, tactile warning and brand identification features can all be enhanced.

Recycling

Growing concern for the environment has led to increased consumer interest in packaging materials and their recyclability.

Metal packaging is 100% recyclable, infinitely, without loss of its inherent properties. It is a permanently available resource that is widely recycled globally and one of the most recycled packaging materials in Europe.

Steel and aluminum are among the most abundant recourses in the world although recycled materials significantly reduce the impact of processing on the environment.

Metal has the best recycling and recovery rates of all key competing packaging materials and is improving year on year:

- Recycling rates in 2010 were 72% for steel cans and 64% for aluminium drinks cans in EU.

- Metal can be recycled an infinite number of times without loss of its essential properties.

- Metal packaging collection for recycling is simple and cost efficient: cans are retrieved easily by any one of the multiple collection systems used across Europe.

- Recycling reduces energy consumption and CO_2 emissions significantly.

Precise engineered features such as beading patterns and end panel designs have enabled light-weighting of the can by up to 35% over the last 20 years adding to its environmental credentials. Weight reduction will continue to be a feature of metal packaging in the years to come.

Metal packaging ranges from tin biscuit containers and aluminum to steel beverage cans. Metal is useful because it's durable, doesn't cost too much and is non-toxic, making it highly suitable for storing food. However, this kind of packaging also has a few disadvantages. Problems with metal packaging vary according to the specific type of metal used.

Corrosion

Some types of metal packaging, such as steel, are vulnerable to the effects of corrosion, which can cause the metal to deteriorate. Corrosion takes place as the metal begins to transform back into its original state; for example, steel turns back into the iron ore it came from. Corrosion is caused by oxidation, brought about when the metal is exposed to air and water. One example of corrosion is rust, which occurs on steel packaging and causes it to flake away. Metal packaging is typically coated in other materials, such as chromium, to prevent corrosion from occurring.

Can't See Contents

Metal packaging may keep a container's contents secure and fresh, but it does pose a disadvantage in that it is not transparent, and so consumers can't see into the packaging to check the contents or to further inspect a potential purchase. This limits the uses for metal packaging within the retail sector, since other packaging materials—such as plastic—are better in some situations. For example, the plastic used in a blister pack used to store nails allows consumers to check the size and type of nails inside the packaging, which wouldn't be possible with metal packaging.

Storage Issues

Tin is often used for certain types of containers, including those for biscuits. Becuase the metal packaging isn't easily bent or squashed by hand, the containers are difficult to store effectively, both during and after use. On the other hand, a paper or plastic container might be easier to fold up or squash and tuck away in a cupboard or other storage facility.

Aluminum and Acidity

Aluminum is another common choice for metal packaging. While aluminum is impervious to corrosion when used to store food products, it does have an issue with acidic foods such as rhubarb and tomatoes. These foods are especially acidic and can be affected by aluminum if the metal is used to store them. The result of using aluminum packaging for these foods is that the food will end up tasting of aluminum.

Metal Packaging Products

Aluminium Foil

Aluminum foil is made by rolling pure aluminum metal into very thin sheets, followed by annealing to achieve dead-folding properties (a crease or fold made in the film will stay in place), which allows it to be folded tightly. Moreover, aluminum foil is available in a wide range of thicknesses, with thinner foils used to wrap food and thicker foils used for trays. Like all aluminum packaging, foil provides an excellent barrier to moisture, air, odors, light, and microorganisms. It is inert to acidic foods and does not require lacquer or other protection. Although aluminum is easily recyclable, foils cannot be made from recycled aluminum without pinhole formation in the thin sheets.

Here are some notable benefits of using aluminium in food packaging.

- Helps in storing food for long periods

    Because of its brilliant impermeability to water vapour and water, aluminium foils help to store food for prolonged time periods. Moreover, because of this wonderful property, aluminium helps to maintain the freshness of foods over long time periods.

- Provides barrier against heat and light

  Because of its excellent ability to act as a barrier against heat and light, heat being one of the main reasons for food decay, aluminium is widely used in food packaging. This is one of the foremost properties that makes aluminium the most preferred element for packaging and storage of foods.

- Packaging is easy

  A rather conspicuous benefit of using aluminium foils for food packaging is that it is easy to pack foods in an aluminium foil. All you need to do is wrap your food with the foil and make sure the wrapping does not squeeze the food packed inside.

- Prevents germs and bacteria

  It isn't just heat that is responsible for deterioration in the quality of food. Aluminium foils prevent foods from coming in contact with bacteria and germs. It is highly resistant to germs and helps keep the food safe from the influence of harmful bacteria.

Aluminum Can

Aluminum cans provide long-term food quality preservation benefits. Aluminum cans deliver 100 percent protection against oxygen, light, moisture and other contaminants. They do not rust, are resistant to corrosion and provide one of the longest shelf lives of any type of packaging. Aluminum-based food canning has an unparalleled safety record. Tamper-resistant and tamper-evident packaging provides consumers with peace of mind that their products have been safely prepared and delivered. A vast variety of products are packaged using aluminum in addition to food and beverages: aerosol products, paint and thousands of other items in the consumer products market.

Beverage Can Technology Continues to Improve

In the past half-century, aluminum beverage can manufacturers have lightened the package by reducing the gauge required to fabricate both the cans and their ends. The first generation of aluminum cans weighed approximately 3 ounces per unit. Today's cans weigh less than half an ounce. Aluminum cans bring packaging benefits as well. They are easily formed, resist corrosion and will not rust. Cans made from aluminum easily support the carbonation pressure required to package soda and withstand pressures of up to 90 pounds per square inch. Believe it or not, four six packs can support a 2-ton vehicle.

Looking Forward: Aluminum Can Market Growth

Many new specialty drinks, particularly energy drinks, are delivered in 8.2-ounce cans. This move was propelled by the wildly successful Red Bull™ brand of energy drink, introduced in the U.S. market in 1997. Since then, the market has swelled with more than 30 new brands following Red Bull's™ lead. The microbrew beer canning market is also growing rapidly. Craft beer packaged in aluminum cans maintains a higher quality. The liquid is not affected by light or exposed to oxygen, which can negatively impact taste. Cans are also portable and infinitely recyclable—they can go where glass can't. Brewers are now positioning the aluminum can as a "mini-keg" that delivers draft-quality fresh beer to the glass.

History of Aluminum Beverage Cans

The modern aluminum beverage can traces its origins to 1959, when Coors introduced the first all-aluminum, seamless, two-piece beverage container. Recycling was instituted at the same time (Coors paid 1 cent for each can returned to the brewery). Aluminum cans made inroads into the soft drink market in 1964, when Royal Crown Cola released both its RC Cola and Diet Rite beverages in two-piece, 12-ounce aluminum containers. In their first year on the market, 1 million cases of soda were packaged using aluminum cans. In addition to being lighter in weight than their steel predecessors, aluminum cans provide a superior surface on which to print text and graphics. This capability increased the opportunity to establish and promote shelf presence and brand awareness.

Important Facts

- The most sustainable beverage container

  As the most valuable package in the bin, aluminum cans are, by far, the most recycled beverage container. The average can contains 70 percent recycled metal.

- Shipping efficiency: the weight advantage

  Aluminum cans are lightweight and easily stacked. This provides storage and shipping efficiencies and limits overall transportation carbon emissions through logistics and supply chains.

- True closed loop recycling

Aluminum cans are recycled over and over again in a true "closed loop" recycling process. Glass and plastic are typically "down-cycled" into products like carpet fiber or landfill liner.

- New markets continue to develop

Nearly 500 craft beer brewers use aluminum to can more than 1,700 different beers. Protection from light and oxygen are two key benefits in addition to the unparalleled sustainability of aluminum packaging.

Metallised Film

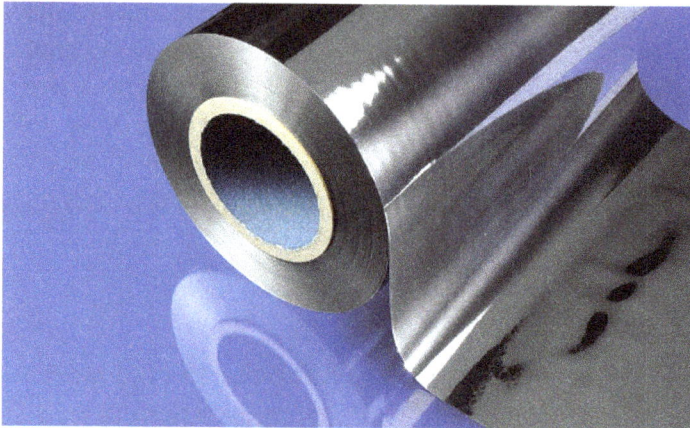

Metallized PET foils are plastic foils which are only evaporated with aluminium powder in a high vacuum. For this process, it is used pure aluminium in powder form with an aluminium content of at least 99,98%. The films produced this way obtain a metallic shine and thereby achieve a special optical effect.

In addition to the optical effect, the aluminum vaporization serves in particular to improve its barrier effects against vapours, gases and aromas in comparison to non-metallized plastic film. Overall, the metallized PET films are less gas-tight than "real" aluminum foil, but more tear resistant, more flexible and lighter. Also, they are less sensitive to corrosion and not as winkle-prone as pure aluminium foil.

PET-met films are made of only 1% of the aluminum that would have been needed for an aluminum foil. Therefore, PET-met is especially interesting as a cost-effective and more environmentally friendly alternative to aluminum.

Metallized PET films are an ideal solution for all applications that require the preservation of flavor and freshness. It is suitable for many applications in the packaging industry as well as for applications in the technical and industrial sectors.

PET films have particular advantages where the higher gas tightness of the aluminum foil is not required.

Metallized PET film can be printed like aluminum foil, and it results in a variety of optical variants. The gloss level of metallized PET film is much higher than for aluminum foil, so with metallized PET film you can achieve a stronger mirror effect. This is also used when printing – for example, without a white primer, yellow colour shimmers gold.

Therefore, PET met films offer an ideal opportunity to combine certain protective and barrier functions with an attractive appearance – even if the protection of the food is not quite as high as with real aluminum foil and the shelf life may need to be shortened.

In combination with other materials, for example, PET met foils are ideal for snacks packaging, thermoformable container covers such as yogurt cups or refill bags for cosmetic products. Coffee and pet kibbles remain dry in those PET packaging and retain their full aroma and flavor for a long time.

In the particular case of acidic, alkaline or salt-containing foods, metallized PET films are a perfect substitute for aluminium foils, as a longer exposure or a large contact area of aluminium with those materials can release potentially harmful emission of aluminium into the edible content.

Insulation systems, for example for insulation against heat radiation in buildings or against heat conduction in vacuum insulation panels, can also be realized with metallized PET films. In addition, PET met films can be used for decorative adhesive tapes. Labels made from metallized PET-film are highly resistant to heat and chemicals, which ensures a long-time durability.

Difference between Aluminium Foil and Metallised Film

One of the major differences between aluminum foil and metallised film is the price, making it very tempting to stick with the cheaper option. However, the difference between semi-permeability and total protection can be significant depending on what you're packaging.

Differences in the permeability of polyethylene, metallised polyethylene and aluminum foil

	Moisture (g/m². Day)	Oxygen (mL/m².day)	UV Light (%transmittance)
PET film, 12.7	31	465	91
Metallised PET	0.8	1.2	5
Aluminum foil 6	0	0	0

If you plan to store perishable food products and powdered substances in ziplock style bags, we recommend using the semi-permeable metallised film bags as they provide a durable barrier that allows some moisture and gas to pass through, preventing the food products from becoming rancid. For liquid based products such as juices, oils and lotions, we would recommend the aluminum foil bags.

- You can see through PET-met foils, not through aluminum foils.

- If high gas tightness or opacity is important, better use aluminum foil.

- For tomatoes, better use PET-met film instead of aluminium foil.

Beverage Can

The stay-tab opening mechanism characteristic of drinking cans

A beverage can (or drinks can) is a can manufactured to hold a single serving of a beverage. In the United States, the can is most often made of aluminum (almost entirely), but cans made in Europe and Asia are an alloy of approximately 55 percent steel and 45 percent aluminum. Aluminum is a widely available, affordable, lightweight metal that is easy to shape. Also, it is far more cost-effective to recycle aluminum than to extract it from its ores.

Fabrication Process

Modern cans are generally produced through a mechanical cold forming process that starts with punching a flat blank from very stiff cold-rolled sheet. This sheet is typically alloy 3104-H19 or 3004-H19, which is aluminum with about one percent manganese and one percent magnesium to give it strength and formability. The flat blank is first formed into a cup about three inches in diameter. This cup is then pushed through a different forming process called "ironing," which forms the can. The bottom of the can is also shaped at this time. The malleable metal deforms into the shape of an open-top can. With the sophisticated technology of the dies and forming machines, the side of the can is significantly thinner than either the top and bottom areas, where stiffness is required. One can-making production line can turn out up to 2400 cans per minute.

Plain lids are stamped out from a coil of aluminum, typically alloy 5182-H49, and are transferred to another press that converts them to easy-open ends. The conversion press forms an integral rivet button in the lid and scores the opening, while concurrently forming the tabs in another die from a separate strip of aluminum. The tab is pushed over the button, which is then flattened to form the rivet that attaches the tab to the lid.

Finally, the top rim of the can is trimmed and pressed inward or "necked" to form a taper conical where the can will later be filled and the lid (usually made of an aluminum alloy with magnesium) attached.

Problems

One problem with the current design is that the top edge of the can may collect dust or dirt in transit, if the can is not packaged in a completely sealed box. Some marketers have experimented with putting a separate foil lid on can tops, and shipping cans in cardboard 12 or 24 pack cases.

Many consumers find the taste of a drink from a can to be different from fountain drinks and those from plastic or glass bottles. In addition, some people believe that aluminum leaching into the fluid contained inside can be dangerous to the drinker's health. The exact role (if any) of aluminum in Alzheimer's disease is still being researched and debated, though the scientific consensus is that aluminum plays no role in development of the disease.

Aluminum cans contain an internal coating to protect the aluminum from the contents. If the internal coating fails, the contents will create a hole and the can will leak in a matter of days. There is some difference in taste, especially noticeable in beer, presumably due to traces of the processing oils used in making the can.

Recycling

In many parts of the world, a deposit can be recovered by turning in empty plastic, glass, and aluminum containers. Unlike glass and plastic, scrap metal dealers often purchase aluminum cans in bulk, even when deposits are not offered. Aluminum is one of the most cost-effective materials to recycle. When recycled without other metals being mixed in, the can/lid combination is perfect for producing new stock for the main part of the can. The loss of magnesium during melting is compensated by the high magnesium content of the lid. Also, the refining of ores such as bauxite into aluminum requires large amounts of electricity, making recycling cheaper than smelting.

Advantantages of Cans

- Save energy: Reusing recycled metals saves as much as 95% of the energy needed to make cans from virgin ores.

- Recyclable: Beverage cans are fully and infinitely recyclable without any quality loss, again and again and again and again.

- Permanent material: The material in cans is only used, not consumed. Because they are infinitely recyclable, metals are a permanent resource.

- Quickly chilled: Beverage cans chill quickly and feel extrafresh to the touch.

- Unbreakable: Unbreakable beverage cans are ideal for large events.

- Material thickness: The sides of today's beverage cans are only 0.065 mm thick – as thin as a human hair. Thanks to ongoing research and development, it is now possible to manufacture cans with far less material than before.

- Light-proof: Beverage cans are absolutely light-proof, protecting the quality of light-sensitive beverages such as beer.

- Stackable: The flat ends and characteristic shape of beverage cans means that a truck carrying cans is able to transport twice as much liquid as a truck loaded with bottles.

- Hermetic seal: Being absolutely airtight, beverage cans keep oxygen out and fizz in, allowing beverages to stay fresh for longer.

- Lightweight: Light and convenient, beverage cans are great for refreshments on the way.

- Fresh: The characteristic sound of a can opening is a unique indicator that the drink inside is absolutely fresh.

Glass Packaging Materials

The two main types of glass container used in food packaging are bottles, which have narrow necks, and jars and pots, which have wide openings. Glass closures are not common today, but were once popular as screw action stoppers with rubber washers and sprung metal fittings for pressurized bottles, e.g. for carbonated beverages, and vacuumized jars, e.g. for heat preserved fruits and vegetables. Ground glass friction fitting stoppers were used for storage jars, e.g. for confectionery.

Glass Containers Market Sectors for Foods and Drinks

A wide range of foods is packed in glass containers. Examples are as follows: instant coffee, dry mixes, spices, processed baby foods, dairy products, sugar preserves, spreads, syrups, processed fruit, vegetables, fish and meat products, mustards and condiments etc. Glass bottles are widely used for beers, wines, spirits, liqueurs, soft drinks and mineral waters. Within these categories of food and drinks, the products range from dry powders and granules to liquids, some of which are carbonated and packed under pressure, and products which are heat sterilized.

The glass package has a modern profile with distinct advantages, including:

- Quality image – consumer research by brand owners has consistently indicated that consumers attach a high quality perception to glass packaged products and they are prepared to pay a premium for them, for specific products such as spirits and liqueurs.

- Transparency – it is a distinct advantage for the purchaser to be able to see the product in many cases, e.g. processed fruit and vegetables.

- Surface texture – whilst most glass is produced with a smooth surface, other possibilities also exist, for example, for an overall roughened ice-like effect or specific surface designs on the surface, such as text or coats of arms. These effects emanate from the moulding but subsequent acid etch treatment is another option.

- Color – as indicated, a range of colors are possible based on choice of raw materials. Facilities exist for producing smaller quantities of nonmainstream colors, e.g. Stolzle's feeder color system.

- Decorative possibilities, including ceramic printing, powder coating, coloured and plain printed plastic sleeving and a range of labelling options.

- Impermeability – for all practical purposes in connection with the packaging of food, glass is impermeable.

- Chemical integrity – glass is chemically resistant to all food products, both liquid and solid. It is odourless.

- Design potential – distinctive shapes are often used to enhance product and brand recognition.

- Heat processable – glass is thermally stable, which makes it suitable for the hot-filling and the in-container heat sterilization and pasteurization of food products.

- Microwaveable – glass is open to microwave penetration and food can be reheated in the container. Removal of the closures is recommended, as a safety measure, before heating commences, although the closure can be left loosely applied to prevent splashing in the microwave oven. Developments are in hand to ensure that the closure releases even when not initially slackened.

- Tamper evident – glass is resistant to penetration by syringes. Container closures can be readily tamper-evidenced by the application of shrinkable plastic sleeves or in-built tamper evident bands. Glass can quite readily accept preformed metal and roll-on metal closures, which also provide enhanced tamper evidence.

- Ease of opening – the rigidity of the container offers improved ease of opening and reduces the risk of closure misalignment compared with plastic containers, although it is recognized that vacuum packed food products can be difficult to open. Technology in the development of lubricants in closure seals, improved application of glass surface treatments together with improved control of filling and retorting all combine to reduce the difficulty of closure removal. However, it is essential in order to maintain shelf life that sufficient closure torque is retained, to ensure vacuum retention with no closure back-off during processing and distribution.

- UV protection – amber glass offers UV protection to the product and, in some cases, green glass can offer partial UV protection.

- Strength – although glass is a brittle material glass containers have high top load strength making them easy to handle during filling and distribution. Whilst the weight factor of glass is unfavourable compared with plastics, considerable savings are to be made in warehousing and distribution costs. Glass containers can withstand high top loading with minimal secondary packaging. Glass is an elastic material and will absorb energy, up to a point, on impact. Impact resistance is improved by an even distribution of glass during container manufacture and subsequent treatment.

- Hygiene – glass surfaces are easily wetted and dried during washing and cleaning prior to filling.

- Environmental benefits – glass containers are returnable, reusable and recyclable. Significant savings in container weight have been achieved by technical advances in design, manufacture and handling.

Glass Pack Integrity and Product Compatibility

- Safety

  Migration studies on glass have shown it to be an inert material as regards its application to packaged foods and, from a health and hygiene viewpoint, it is regarded as an optimal material for containing food and drinks.

- Product compatibility

  Glass containers are noted for the fact that they enable liquid and solid foods to be stored for long periods of time without adverse effects on the quality or flavour of the product.

- Consumer acceptability

  Market research has indicated that consumers attach a high quality perception to glass packaged products. Recent findings of a report on consumer perceptions carried out by The Design Engine, on behalf of Rockware Glass, concluded that there are five key and largely exclusive benefits for food packaging in glass, namely:

  1. Aesthetic appeal.

  2. Quality perception.

  3. Preferred taste.

  4. Product visibility and associated appetite appeal.

  5. Resealability.

Glass and Glass Container Manufacture

Melting

Glass is melted in a furnace at temperatures of around 1350°C (2462° F) and is homogenized in the melting process, producing a bubble-free liquid. The molten glass is then allowed to flow through a temperature controlled channel (fore hearth) to the forming machine, where it arrives via the feeder at the correct temperature to suit the container to be produced. For general containers, suitable for foods and carbonated beverages, this would be in the region of 1100°C (2012° F).

Container Forming

In the feeder the molten glass is extruded through an orifice of known diameter at a predetermined rate and is cropped into a solid cylindrical shape. The cylinder of glass is known in the trade as a *gob* and is equivalent in weight to the container to be produced. The gob is allowed to free-fall through a series of deflectors into the forming machine, also known as the IS or individual section machine, where it enters the parison. The parison com- prises a neck finish mould and a parison mould, mounted in an inverted position. The parison is formed by either pressing or blowing the gob to the shape of the parison mould. The parison is then reinverted, placed into the final mould and blown out to the shape of the final mould, from where it emerges at a temperature of approximately 650°C (1200° F). A container is said to have been produced by either the press and blow or blow and blow process.

In general terms, the press and blow process is used for jars and the blow and blow process for bottles. An alternative, for lightweight bottles, is the narrow neck press and blow process.

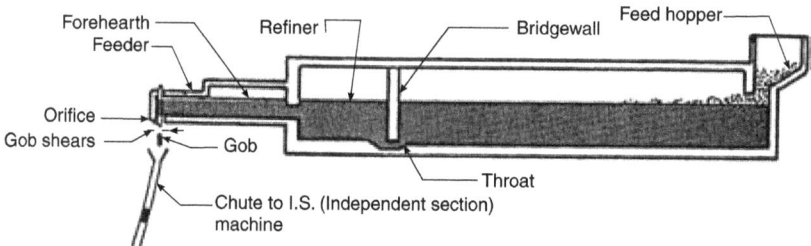

Figure: The feeder – molten glass is extruded through the orifice at a predetermined rate and is cropped into a solid cylinder known as a gob

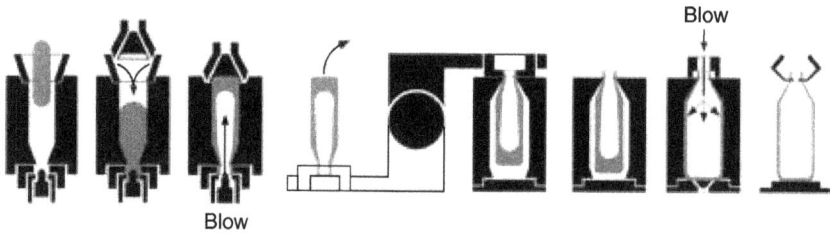

Figure: The blow and blow forming process

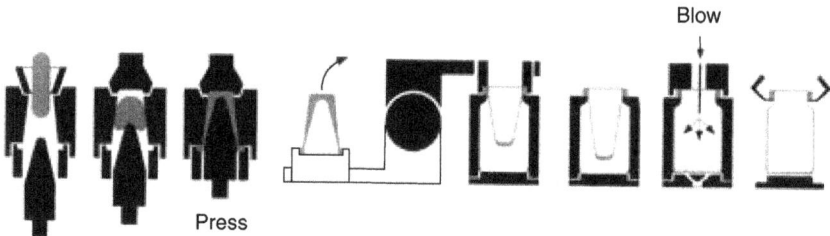

Figure: The wide mouth press and blow forming process

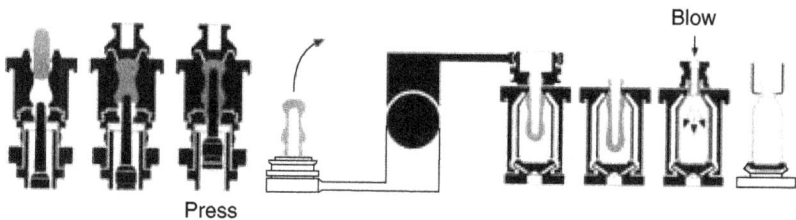

Figure: The narrow neck press and blow forming process

The press and blow process is generally best suited to produce jars with a neck finish size of ≥35 mm (≥1.25''); the other two processes are more suited to produce bottles with a neck finish size of ≤35 mm (≤1.25'').

The narrow neck press and blow process offers better control of the glass distribution than the blow and blow process, allowing weight savings in the region of 30% to be made.

Design Parameters

One of the design parameters to be borne in mind when looking at the functionality of a glass container is that the tilt angle for a wide-mouthed jar should be ≥22° and that for a bottle ≥16°. These parameters are indicative of the least degree of stability that the container can withstand.

Surface Treatments

Once formed, surface treatment is applied to the container in two stages: hot end and cold end treatment, respectively.

Hot End Treatment

The purpose of hot end surface treatment is to prevent surface damage whilst the bottle is still hot and to help maintain the strength of the container.

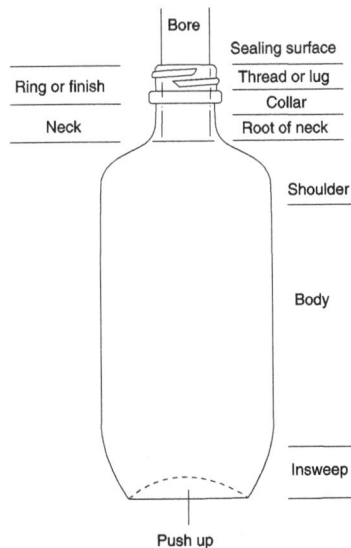

Figure: The parts of a glass container

The most common coating material deposited is tin oxide, although derivatives of titanium are also used. This treatment tends to generate high friction surfaces; to overcome this problem, a lubricant is added.

Cold End Treatment

The second surface treatment is applied once the container has been annealed. Annealing is a process which reduces the residual strain in the container that has been introduced in the forming process. The purpose of the cold end treatment is to create a lubricated surface that does not break down under the influence of pressure or water, and aids the flow of containers through a high speed filling line. Application is by aqueous spray or vapor, care being taken to prevent entry of the spray into the container, the most commonly used lubricants being derivatives of polyester waxes or polyethylene. The surface tension resulting from this treatment can be measured by using Dynes indicating pens.

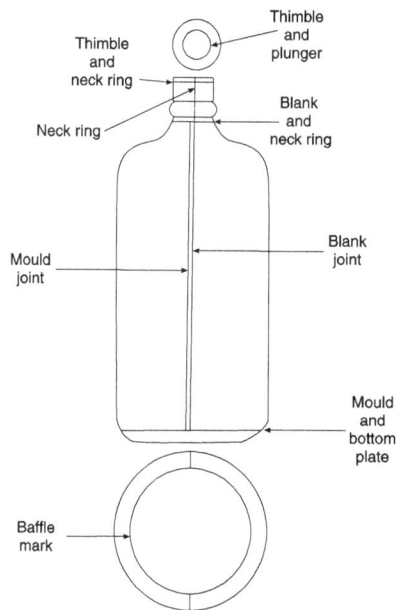

Figure: The positions of the moulding joints of the glass container

Labelling compatibility should be discussed with either the adhesive supplier or the adhesive label supplier depending on the type of label to be used.

Low-cost Production Tooling

The tooling cost for a glass container is approximately one-fifth that of a plastic container. Whilst the numbers produced per cavity are lower than for plastic, this can be advantageous, because the design can be modified or completely revamped in a much shorter time-span than plastic; thus, the product image can be updated and the product marketability kept alive. The numbers produced per mould cavity vary depending on the number of production runs required, the complexity of the shape and the embossing detail. In general, 750 000 pieces can be produced from a complex mould and 1 000 000 pieces from a mould of a simple round shape. There can be upwards of 20 moulds per production set.

Container Inspection and Quality

As with packaging in general, quality assurance is needed to ensure that consumer safety, brand owner's needs and efficiency in handling, packing, distribution and merchandising are achieved.

Quality assurance needs are defined and incorporated into the specification of the glass container at the design stage and by, consistency in manufacture, thereby meeting the needs of packing, distribution and use. Quality control, on the other hand, comprises the procedures, including on-line inspection, sampling and test methods used to control the process and assess conformity with the specification.

The techniques used can broadly be defined as chemical, physical and visual. Chemical testing by spectrophotometry, flame photometry and X-ray fluorescence is used to check raw materials and the finished glass. Small changes in the proportions and purity of raw materials can have a significant effect on processing and physical properties.

Physical tests include checking dimensional tolerances, tests for color, impact strength, thermal shock resistance and internal pressure strength. Visual tests check for defects that can be seen.

The list of possible visually observable defects is quite long and though most of them are comparatively rare, it is essential that production be checked by planned procedures. The categories of these defects comprise various types of cracks, glass strands, foreign bodies and process material contamination from the process environment, misshapes and surface marks of various kinds.

Defects are classified as being,

- Critical, e.g. defects which endanger the consumer or prevent use in packing;

- Major, e.g. defects which seriously affect efficiency in packing;

- Minor, e.g. defects that relate to appearance even though the container is functionally satisfactory.

Visual inspection on manufacturing and packing lines is assisted today by automatic monitoring systems, as shown in figure below, where this is appropriate. Systems are available for container sidewall inspection using multiple cameras that detect opaque and transparent surface defects. Infrared cameras can be used in a system to examine containers directly after formation. On the packing line, claims have been made that foreign bodies can be detected in glass containers running at speeds up to 60 000 capped beer bottles per hour.

An X-ray system such as that from Heimann Systems Corp. is designed for the automatic inspection of products packed in jars.

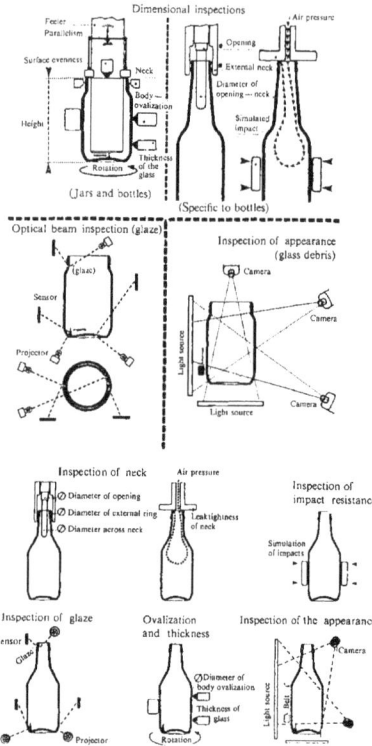

Figure: Principles of production line inspections

The system can detect foreign materials such as ferrous and stainless metals, glass particles, stone, bone and plastic materials. This equipment runs at 100 m min^{-1}.

Closure Selection

Closures for glass packaging containers are usually metal or plastic, though cork is still widely used for wines and spirits. Effecting a seal is achieved either by a tight fitting plug, a screw threaded cap applied with torque in one of several ways or a metal cap applied with pressure and edge crimping. Hermetic or airtight sealing can be achieved by heat sealing a flexible barrier material to the glass usually with an overcap for protection and subsequent reclosing during use. The aluminium foil cap applied to a milk bottle is one of the simplest forms of closure.

All these closures are applied to what is known as the finish of the container. This may seem an odd name for the part of the container which is formed first but in fact this name goes back to the time of blowing and forming glass containers by hand when the rim was the last part to be formed and therefore called the finish.

Four key dimensions determine the finish as shown in figure. Industry- wide standards for these dimensions have been agreed upon. The contour of glass threads are round, and closures, both metal and plastic, with symmetrical threads will fit the appropriate containers.

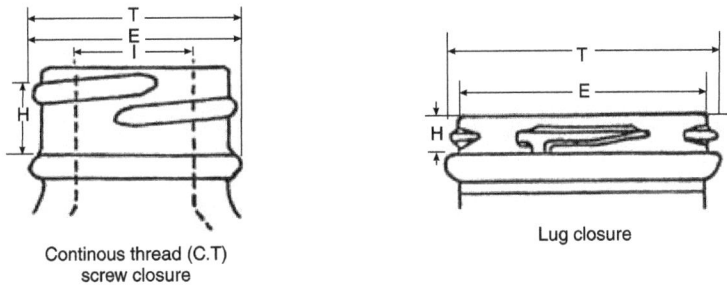

Continous thread (C.T)
screw closure

Lug closure

Figure: Standard finish nomenclature
I: Diameter at smallest opening inside finish, T: Thread diameter measured
across the threads, E: Thread root diameter, H: Top of finish to top of
bead or to intersection with bottle shoulder on beadless designs

Careful choice of closure is essential. Too large a closure can create leakage due to the force generated upon it either from internal gas pressure or from heating during processing of the product. Too small a closure may well intro- duce an interference fit between the minimum through bore on the glass container and the filler tube. The types of closure available fall into three main categories:

1. Normal seal

2. Vacuum seal

3. Pressure seal.

Normal Seals

Normal seals, that is those used for non-vacuum/non-pressure filled products, comprise composite closures of plastic/foil, for products such as coffee, milk powders, powder and granular products in general and for mustards, milk and yoghurts.

Glass lends itself both to induction and conduction sealing without prior treatment of the glass finish, but is considered suitable only for dry powders and products, such as peanut butter and chocolate spreads, which do not require a further heating process.

Crimped seals using foils, used for instance on milk containers, have been used for a long time, and are cost effective.

Vacuum Seals

Vacuum seals are metal closures with a composite liner to seal onto the glass rim. They can be pressed or twisted into place, at which time a vacuum is created by flushing the headspace with steam. They lend themselves quite readily to in-bottle pasteurization and retort sterilization and sizes range from 28 to 82 mm. For beverages, sizes are usually in the 28–40 mm range.

Pressure Seals

Pressure seals can be metal or plastic with a composite liner to make the seal, and can either be pressed or twisted into place. They include:

- Preformed metal, e.g. crown or twist crown.

- Metal closures rolled-on to the thread of the glass.

- Roll-on pilfer proof (ROPP).

- Preformed plastic screwed into position with or without a tamper evidence band.

Selecting the correct glass finish to suit the closure to be used is essential. Advice on suitability should be sought from both the closure and the glass manufacturers before the final choice is made.

Thermal Processing of Glass Packaged Foods

Glass containers lend themselves to in-bottle sterilization and pasteurization for both hot and cold filled products. Subject to the headspace volume conditions being maintained and thermal shock ground rules being observed, no problems will be experienced.

In general terms, hot-fill products filled at 85°C and then cooled will require a minimum headspace of 5%, whilst a cold filled product requiring sterilization at 121°C will require a 6% minimum head space. In all cases the recommendations of the closure supplier should be obtained before preparing the design brief. It should be noted that the thermal shock of glass containers is twice as high when cooling down as when warming up. To avoid thermal shock, cool down differentials should not exceed 40°C and warm up differentials should not exceed 65°C.

Internal pressure resistance: A well-designed glass container can withstand an internal pressure of up to 10 bar, although the norm required rarely exceeds 5 bar. It is also capable of withstanding internal vacuum conditions and filling of thick concentrates, with steam-flushing of the head- space to produce the initial vacuum requirements for the closure seal.

Resealability: Preformed metal, rolled-on metal and preformed plastic closures can all be readily applied to the neck finish of a glass container. Prise off crown closures offer no reseal, whilst the twist-off crown satisfies reseal performance within reason.

Plastic Sleeving and Decorating Possibilities

Glass containers can accept a wide range of decorative formats, i.e. labelling, silk screen printing with ceramic inks, plastic sleeving, acid etching, organic and inorganic color coating and embossing. The rigidity of the container offers a good presentation surface for decorating, which is not subject to distortion from internal pressure or internal vacuum.

When plastic sleeving the container, it is essential to test the sleeving film under in-bottle pasteurization temperatures to ensure that no secondary movement of the sleeve occurs. Care should also be taken not to exceed the stretch limits of the film by ensuring the maximum and minimum diameters of the area to be sleeved do not fall outside the stretch ratios of the film specification.

Sleeving also offers fragment retention properties, should the container become damaged in use.

Strength in Theory and Practice

The theoretical strength of glass can be calculated and it is extremely high. In practice, the strength is much lower due to surface blemishes, such as micro- cracks, which are vulnerable stress points for impacts such as occur during handling and on packing lines. Work has therefore been concentrated on:

- Improving the surface to reduce defects
- Improving surface coatings in manufacture
- Avoiding stress during manufacture and use.

Major investigations of packing line performance, noting all breakages, and using techniques such as high speed video, can lead to improvements in performance by eliminating stress points.

Broken bottles can be reconstructed and thereby demonstrate the type of impact which caused the failure, such as whether it occurred at a slow or fast rate or whether it was caused by an external, or internal, pressure related fault. The strength of a glass container is also dependent on shape and thickness. The inter-relationship can be subjected to computer modeling as a design-aid to:

- Identify vulnerable features in a proposed design
- Calculate the effect of modifying the design
- Simulate the effect of lightweighting by reducing thickness.

Specific tests can be carried out on containers to check:

- Vertical crushing
- Internal pressure
- Thermal shock.

Thermal shock relates to heat transfer and as glass is a very good insulator, heat is conducted slowly across the walls when a hot liquid is filled. Another important heat related property is the dimensional change per degree change in temperature, which is low for glass. This property is also subject to the glass formulation, e.g. Pyrex is a well known type of glass with an even lower heat expansion compared with standard white flint soda glass. This is achieved by replacing some of the soda with boric oxide and increasing the proportion of silica.

It has been recognized that achieving an even distribution of glass in the walls of a container is the major factor in successfully reducing weight whilst maintaining adequate strength.

Glass Pack Design and Specification

Concept and Bottle Design

Leading glass manufacturers have state-of-the-art design expertise and systems that can be readily integrated with design house concepts to design a container which meets the requirements of branding, manufacture, filling and distribution under recommended good manufacturing practices and procedures.

The brand manager/packaging technologist can now quite readily bring together all the expertise necessary to produce a food container of the ultimate design, cost and quality to meet all their needs.

An understanding of the product specification and the filling line requirements is essential at the concept design stage. The information required includes:

- Type and density of the product
- Carbonation level, if required, in the product
- Closure type/neck specification required
- Quantity to be filled
- Type of filling process (hot-fill/cooled, hot-fill/pasteurized, ambient-fill/ sterilized or any combination of these processes)
- Is the container to be a specified volume-measuring container?
- What type of filler is to be used (volumetric or vacuum-assisted)?
- What is the filler tube size/diameter?
- Is the container to be refillable or single-trip?
- Speed of the filling operation, i.e. bottles per minute
- Impact forces on the process line
- What pallet size is to be used in the distribution of filled stock?

- Is the depalletizer operation sweep-off or lift-off?

From this information, the glass manufacturer can select the correct finish and closure design, surface treatment requirements, the type of pack to be used for distribution to the filling line and the handling systems. Wherever possible, the body size of the container should ensure a snug fit to the pallet, since any over- hang of the glass beyond the edge of the pallet could result in breakage in transit, whilst underhang on the pallet could lead to instability. Compression, tension strapped packs can be accommodated together with live bed deliveries. This creates a highly efficient delivery system with minimal stock-holding on site, by means of just-in-time (JIT) deliveries.

Ever more challenging briefs are demanding more from packaging materials. It is well known that consumers have an innate, high quality perception of glass packaging. This emotional connection between consumer and brand is highly valued by food and drink consumers. Add to this the ability of glass to be formed into unique shapes with a wide range of decoration techniques and it is clear why glass is also the preferred choice for designers. With increased emphasis on production speed and efficiency, design freedom decreases.

Low volume: For low volume, limited or special edition products, the design freedom is high, as hand operated or semi-automated processing lines are used. Bottles may be produced using single gob machines and have a high (+0.8) capacity to weight ratio.

Main stream: For main stream production volumes, design freedom decreases, with automatic filling lines and bulk distribution being important. Bottles will be produced using larger double gob machines and have capacity to weight ratios of around 0.6–0.7.

High volume: For high volume brands, which probably have multinational distribution, the design freedom is strictly controlled to ensure compatibility with very high speed (+1000 bpm) filling lines. These brands will be produced using the largest double and triple gob NNPB machines and have capacity to weight ratios down to 0.5. A *full circle* design process creates a range of radical design options and a common sense view on the likely costs and implications of each concept. This ensures all design options are fully explored and the best design solutions are rapidly brought to the market.

Concept design: A concept design team focuses on the packaging as a brand communication tool. Using brand analysis they ensure the pack is as active as possible, at the point of sale and in use, in communicating the brand's value and positioning. Concept designers are able to work very closely with a customer's design agency, supporting the design process so that a wide range of creative options are explored, yet at the same time highlighting the practical consequences of the design options. This allows realistic, balanced decisions to be made at the earliest possible stage of the project.

Product design: Taking computer information from the concept designers, or any design agent, product designers apply a series of objective tests to the design to ensure it is fit for the purpose. These include stress analysis to check retention of carbonated products, packing line stability, and impact analysis to assess the containers' filling line performance. Strength for stacking and distribution is also checked. On completion of these tests a detailed specification of the design is issued and a 3D computer model displayed. The 3D computer model is used to create exact models for market research and to seek approval for new designs.

Mould design: The mould design team translates the product specification into mould equipment that will reproduce the container millions of times. Depending on the manufacturing plant and the process to be used, the mould equipment will vary. The level of precision required for modern glass container production is extremely exacting and has a direct effect on product quality.

The product design computer model is used to control all aspects of the design, ensuring exact replication of the design into glass. The design is now ready to be transferred to the mould makers.

Production: Quality information from each production run is fed back to the product and mould design teams to ensure best practice is used on all designs and that design teams are up to date with improvements in manufacturing capability. This closes the full circle.

Packing – Due Diligence in the use of Glass Containers

Receipt of deliveries: Glass containers are usually delivered on bulk palletized shrink-wrapped pallets. A check should be made for holes in the pallet shroud and broken glass on the pallet, and any damaged pallets rejected. The advice note should be signed accordingly, informing the supplier and returning the damaged goods.

Storage/on-site warehousing: Pallets of glass must not be stored more than six high, they must be handled with care and not shunted. Fork-lift trucks should be guarded to prevent the lift masts contacting the glass. Where air rinser cleaning is used on the filling line, the empty glass containers should not be stored outside. Pallets damaged in on-site warehousing must not be forwarded to the filling area until they have been cleared of broken glass.

Depalletization: A record should be made of the sequence and time of use of each pallet and the product batch code. Plastic shrouds must be removed with care to prevent damage to the glass; if knives are used, the blade should be shrouded at all times, so as not to damage the glass. It is necessary to ensure that the layer pads between the glass containers are removed in such a way as to prevent any debris present from dropping onto the next layer of glass. Breakages must be recorded and clean-up equipment provided to prevent any further contamination.

Cleaning Operation

Air rinse: The glass must be temperature-conditioned to prevent condensate forming on the inside, which would inhibit the removal of card- board debris. The air pressure should be monitored to ensure that debris is not suspended and allowed to settle back into the container.

On-line water rinse: Where hot-filling of the product takes place, it is essential to ensure that the temperature of the water is adequate to prevent thermal shock at the filler, i.e. not more than 60°C (140° F) differential.

Returnable wash systems: The washer feed area must be checked to ensure that the bottles enter the washer cups cleanly. A washer-full of bottles must not be left soaking overnight. In the longer term this would considerably weaken the container and could well create a reaction on the bottle surface between the hot end coating and the caustic in the washer. Where hot-filling is taking place, it is necessary to ensure that the correct temperature is reached to prevent thermal shock at the filler.

Filling operation: Clean-up instructions should be issued and displayed, so that the filling line crew know the procedure to follow should a glass container breakage occur and the need to record all breakages. It is essential to ensure that flood rinsing of the filler head in question is adequate to prevent contamination of further bottles. It is necessary to ensure that filling levels in the container comply with trading standards' requirements for measuring containers.

Capping: Clean-up instructions on the procedure to follow should breakage occur in the capper should be issued and displayed, and all breakages recorded. The application torque of the caps and vacuum levels must be checked at prescribed intervals, as must the cap security of carbonated products.

Pasteurization/sterilization: It is necessary to ensure that cooling water in the pasteurizer or sterilization retort does not exceed a differential of more than 40°C (104° F), to prevent thermal shock situations. The ideal temperature of the container after cooling is 40°C, which allows further drying of the closure and helps prevent rusting of metal closures. Air knives should be used to remove water from closures to further minimize the risk of rusting.

Labelling: Where self-adhesive labels are to be used, all traces of condensate must be eliminated to obtain the optimum conditions for label application. Adhesives must not be changed without informing the glass supplier, since this could affect the specification of adhesives/surface treatments.

Distribution: It is essential to ensure that the arrangement of the glass containers in the tray, usually plastic or corrugated fibreboard, is adequate to prevent undue movement during distribution, that the shrink-wrap is tight and that the batch coding is correct and visible.

Warehousing: The pallets of filled product must be carefully stacked to prevent isolated pockets of high loading that might create cut through in the lining compound of the container closures, as this would result in pack failures.

Quality management: The procedures of good management practice in the development, manufacture, filling, closing, processing, storage and distribution of food products in glass containers discussed in this chapter have been developed to ensure that product quality and hygiene standards are achieved along with consumer and product safety needs. Their application indicates due diligence in meeting these needs. It is essential that all procedures are clearly laid down, training is provided in their use and that regular checks are made on their implementation.

Companies can demonstrate due diligence by achieving certification under an accepted Quality Management Standard, such as ISO 9000. In the UK, the British Retail Consortium (BRC) and The Institute of Packaging (IOP) have cooperated in the publication of a Technical Standard and Protocol, which can be integrated with their ISO 9000 procedures. The BRC is a trade association representing around 90% of the retail trade and the IOP is the professional membership body, established in 1947, for the packaging industry. The IOP has amongst its objectives the education and training of people engaged in the packaging industry. This technical standard and protocol requires companies to:

- Adopt a formal Hazard Analysis System

- Implement a documented Technical Management System

- Define and control factory standards, product and process specifications and personnel needs.

Environmental Profile

Reuse

Glass containers can be reused for food use. However, there is only one well established household example in the UK – that of the daily doorstep delivery of fresh milk in bottles and the collection of the empty bottles. There are wide disparities in the number of trips that can be expected depending on the location, with around 12 trips per bottle being the national average. The decline of doorstep delivery has been rapid over the last decade but the system of reuse is well established. In the licensed trade, and in most places where drinks are served to customers, the drinks manufacturers operate returnable systems.

Recycling

Glass is one of the easiest materials to be recycled because it can be crushed, melted and reformed an infinite number of times with no deterioration of structure. It is the only packaging material that retains all its quality characteristics when it is recycled.

Using recycled glass, in place of virgin raw materials, to manufacture new glass containers reduces:

- The need to quarry and transport raw materials

- The energy required to melt the glass

- Furnace chimney emissions

- The amount of solid waste going into landfill.

In order to recycle glass, it must first be recovered. In the UK, glass is brought by consumers to bottle banks. Currently, approximately 600 000 t/a are recovered – a figure which must increase sharply if the UK is to meet increased European Union targets for glass recovery at currently generated levels of glass container waste. Currently the recycled content of the average glass container is around 33%. The recycled proportion is higher for green than for clear containers and reflects the proportions of clear and green glass taken to bottle banks by the public. Green glass may now contain as much as 85–90% recycled glass.

Reduction – Lightweighting

In the period 1992–2002, it is claimed that the average weight of glass containers has been reduced by 40–50%. This is an average reduction. Some brand owners still retain heavy containers, e.g. spirits and liqueurs, and this causes the progress made by the glass industry to reduce the weight of packaging to be understated.

Glass Packaging Products

Plastic Packaging Materials

Plastics are made by condensation polymerization or addition polymerization (polyaddition) of monomer units. In polycondensation, the polymer chain grows by condensation reactions between molecules and is accompanied by formation of low molecular weight byproducts such as water and methanol. Polycondensation involves monomers with at least 2 functional groups such as alcohol, amine, or carboxylic groups. In polyaddition, polymer chains grow by addition reactions, in which 2 or more molecules combine to form a larger molecule without liberation of by-products. Polyaddition involves unsaturated monomers; double or triple bonds are broken to link monomer chains. There are several advantages to using plastics for food packaging. Fluid and moldable, plastics can be made into sheets, shapes, and structures, offering considerable design flexibility. Because they are chemically resistant, plastics are inexpensive and lightweight with a wide range of physical and optical properties. In fact, many plastics are heat sealable, easy to print, and can be integrated into production processes where the package is formed, filled, and sealed in the same production line. The major disadvantage of plastics is their variable permeability to light, gases, vapors, and low molecular weight molecules.

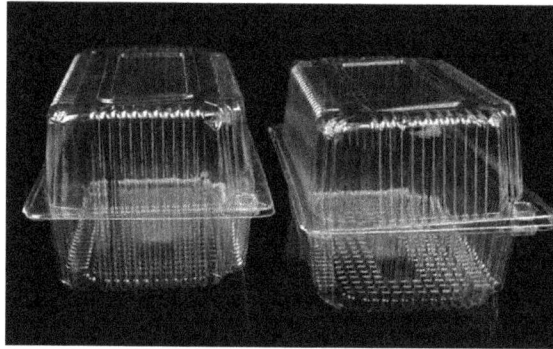

There are 2 major categories of plastics: thermosets and thermoplastics. Thermosets are polymers that solidify or set irreversibly when heated and cannot be remolded. Because they are strong and durable, they tend to be used primarily in automobiles and construction applications such as adhesives and coatings, not in food packaging applications. On the other hand, thermoplastics are polymers that soften upon exposure to heat and return to their original condition at room temperature. Because thermoplastics can easily be shaped and molded into various products such as bottles, jugs, and plastic films, they are ideal for food packaging. Moreover, virtually all thermoplastics are recyclable (melted and reused as raw materials for production of new products), although separation poses some practical limitations for certain products. The recycling process requires separation by resin type as identified by the American Plastics Council.

Resin	Code	Amount generated (thousand tons)	Amount recycled (thousand tons)
Polyethylene terephthalate	1	2860	540
High-density polyethylene	2	5890	520
Polyvinyl chloride	3	1640	
Low-density polyethylene	4	6450	190[a]
Polypropylene	5	4000	10
Polystyrene	6	2590	
Other resins	7	5480	390

Source: American Plastics Council (2006b) and EPA (2006a).
[a]Includes linear low-density polyethylene.

Table: Resin identification codes for plastic recycling

There have been some health concerns regarding residual monomer and components in plastics, including stabilizers, plasticizers, and condensation components such as bisphenol A. Some of these concerns are based on studies using very high intake levels; others have no scientific basis. To ensure public safety, FDA carefully reviews and regulates substances used to make plastics and other packaging materials. Any substance that can reasonably be expected to migrate into food is classified as an indirect food additive subject to FDA regulations. A threshold of regulation—defined as a specific level of dietary exposure that typically induces toxic effects and therefore poses negligible safety concerns may be used to exempt substances used in food contact materials from regulation as food additives. FDA revisits the threshold level if new scientific information raises concerns. Furthermore, FDA advises consumers to use plastics for intended purposes in accordance with the manufacturer's directions to avoid unintentional safety concerns.

Despite these safety concerns, the use of plastics in food packaging has continued to increase due to the low cost of materials and functional advantages over traditional materials such as glass and tinplate. Multiple types of plastics are being used as materials for packaging food, including polyolefin, polyester, polyvinyl chloride, polyvinylidene chloride, polystyrene, polyamide, and ethylene vinyl alcohol. Although more than 30 types of plastics have been used as packaging materials, polyolefins and polyesters are the most common.

Polyolefins: Polyolefin is a collective term for polyethylene and polypropylene, the 2 most widely used plastics in food packaging, and other less popular olefin polymers. Polyethylene and polypropylene both possess a successful combination of properties, including flexibility, strength, lightness, stability, moisture and chemical resistance, and easy processability, and are well suited for recycling and reuse.

The simplest and most inexpensive plastic made by addition polymerization of ethylene is polyethylene. There are 2 basic categories of polyethylene: high density and low density. High-density polyethylene is stiff, strong, tough, resistant to chemicals and moisture, permeable to gas, easy to process, and easy to form. It is used to make bottles for milk, juice, and water; cereal box liners; margarine tubs; and grocery, trash, and retail bags. Low-density polyethylene is flexible, strong, tough, easy to seal, and resistant to moisture. Because low-density polyethylene is relatively transparent, it is predominately used in film applications and in applications where heat sealing is necessary. Bread and frozen food bags, flexible lids, and squeezable food bottles are examples of low-density polyethylene. Polyethylene bags are sometimes reused. Of the 2 categories of polyethylene, high-density polyethylene containers, especially milk bottles, are the most recycled among plastic packages.

Harder, denser, and more transparent than polyethylene, polypropylene has good resistance to chemicals and is effective at barring water vapor. Its high melting point (160°C) makes it suitable for applications where thermal resistance is required, such as hot-filled and microwavable packaging. Popular uses include yogurt containers and margarine tubs. When used in combination with an oxygen barrier such as ethylene vinyl alcohol or polyvinylidene chloride, polypropylene provides the strength and moisture barrier for catsup and salad dressing bottles.

Polyesters: Polyethylene terephthalate (PET or PETE), polycarbonate, and polyethylene naphthalate (PEN) are polyesters, which are condensation polymers formed from ester monomers that result from the reaction between carboxylic acid and alcohol. The most commonly used polyester in food packaging is PETE.

Polyethylene terephthalate: Formed when terephthalic acid reacts with ethylene glycol, PETE provides a good barrier to gases (oxygen and carbon dioxide) and moisture. It also has good resistance to heat, mineral oils, solvents, and acids, but not to bases. Consequently, PETE is becoming the packaging material of choice for many food products, particularly beverages and mineral waters. The use of PETE to make plastic bottles for carbonated drinks is increasing steadily (van Willige and others 2002). The main reasons for its popularity are its glass-like transparency, adequate gas barrier for retention of carbonation, light weight, and shatter resistance. The 3 major packaging applications of PETE are containers (bottles, jars, and tubs), semirigid sheets for thermoforming (trays and blisters), and thin-oriented films (bags and snack food wrappers). PETE exists both as an amorphous (transparent) and a semicrystalline (opaque and white) thermoplastic material. Amorphous PETE has better ductility but less stiffness and hardness than semicrystalline PETE, which has good strength, ductility, stiffness, and hardness. Recycled PETE from soda bottles is used as fibers, insulation, and other nonfood packaging applications.

Polycarbonate: Polycarbonate is formed by polymerization of a sodium salt of bisphenol acid with carbonyl dichloride (phosgene). Clear, heat resistant, and durable, it is mainly used as a replacement for glass in items such as large returnable/refillable water bottles and sterilizable baby bottles. Care must be taken when cleaning polycarbonate because using harsh detergents such as sodium hypochlorite is not recommended because they catalyze the release of bisphenol A, a potential health hazard. An extensive literature analysis by vom Saal and Hughes (2005) suggests the need for a new risk assessment for the low-dose effects of this compound.

Polyethylene naphthalate: PEN is a condensation polymer of dimethyl naphthalene dicarboxylate and ethylene glycol. It is a relatively new member of the polyester family with excellent performance because of its high glass transition temperature. PEN's barrier properties for carbon dioxide, oxygen, and water vapor are superior to those of PETE, and PEN provides better performance at high temperatures, allowing hot refills, rewashing, and reuse. However, PEN costs 3 to 4 times more than PETE. Because PEN provides protection against transfer of flavors and odors, it is well suited for manufacturing bottles for beverages such as beer.

Polyvinyl chlorid: Polyvinyl chloride (PVC), an addition polymer of vinyl chloride, is heavy, stiff, ductile, and a medium strong, amorphous, transparent material. It has excellent resistance to chemicals (acids and bases), grease, and oil; good flow characteristics; and stable electrical properties. Although PVC is primarily used in medical and other nonfood applications, its food uses include bottles and packaging films. Because it is easily thermoformed, PVC sheets are widely used for blister packs such as those for meat products and unit dose pharmaceutical packaging.

PVC can be transformed into materials with a wide range of flexibility with the addition of plasticizers such as phthalates, adipates, citrates, and phosphates. Phthalates are mainly used in nonfood packaging applications such as cosmetics, toys, and medical devices. Safety concerns have emerged over the use of phthalates in certain products, such as toys. Because of these safety concerns, phthalates are not used in food packaging materials in the United States (HHS 2005); instead, alternative nonphthalate plasticizers such as adipates are used. For example, di-(2-ethylhexyl) adipate (DEHA) is used in the manufacture of plastic cling wraps. These alternative plasticizers also have the potential to leach into food but at lower levels than phthalates. Low levels of DEHA have shown no toxicity in animals. Finally, PVC is difficult to recycle because it is used for such a variety of products, which makes it difficult to identify and separate. In addition, incineration of PVC presents environmental problems because of its chlorine content.

Polyvinylidene chloride: Polyvinylidene chloride (PVDC) is an addition polymer of vinylidene chloride. It is heat sealable and serves as an excellent barrier to water vapor, gases, and fatty and oily products. It is used in flexible packaging as a monolayer film, a coating, or part of a co-extruded product. Major applications include packaging of poultry, cured meats, cheese, snack foods, tea, coffee, and confectionary. It is also used in hot filling, retorting, low-temperature storage, and modified atmosphere packaging. PVdC contains twice the amount of chlorine as PVC and therefore also presents problems with incineration.

Polystyrene: Polystyrene, an addition polymer of styrene, is clear, hard, and brittle with a relatively low melting point. It can be mono-extruded, co-extruded with other plastics, injection molded, or foamed to produce a range of products. Foaming produces an opaque, rigid, lightweight material with impact protection and thermal insulation properties. Typical applications include protective packaging such as egg cartons, containers, disposable plastic silverware, lids, cups, plates, bottles, and food trays. In expanded form, polystyrene is used for nonfood packaging and cushioning, and it can be recycled or incinerated.

Polyamide: Commonly known as nylon (a brand name for a range of products produced by DuPont), polyamides were originally used in textiles. Formed by a condensation reaction between diamine and diacid, polyamides are polymers in which the repeating units are held together by amide links. Different types of polyamides are characterized by a number that relates to the number of carbons in the originating monomer. For example, nylon-6 has 6 carbons and is typically used in packaging. It has mechanical and thermal properties similar to PETE, so it has similar usefulness, such as boil-in bag packaging. Nylon also offers good chemical resistance, toughness, and low gas permeability.

Ethylene vinyl alcohol: Ethylene vinyl alcohol (EVOH) is a copolymer of ethylene and vinyl alcohol. It is an excellent barrier to oil, fat, and oxygen. However, EVOH is moisture sensitive and is thus mostly used in multilayered co-extruded films in situation where it is not in direct contact with liquids.

Laminates and co-extrusions: Plastic materials can be manufactured either as a single film or as a combination of more than 1 plastic. There are 2 ways of combining plastics: lamination and co-extrusion. Lamination involves bonding together 2 or more plastics or bonding plastic to another material such as paper or aluminum (as discussed in the section on metal). Bonding is commonly achieved by use of water-, solvent-, or solids-based adhesives. After the adhesives are applied to 1 film, 2 films are passed between rollers to pressure bond them together. Lamination using laser rather than adhesives has also been used for thermoplastics. Lamination enables reverse printing, in which the printing is buried between layers and thus not subject to abrasion, and can add or enhance heat sealability.

In co-extrusion, 2 or more layers of molten plastics are combined during the film manufacture. This process is more rapid (requires 1 step in comparison to multiple steps with lamination) but requires materials that have thermal characteristics that allow co-extrusion. Because co-extrusion and lamination combine multiple materials, recycling is complicated. However, combining materials results in the additive advantage of properties from each individual material and often reduces the total amount of packaging material required. Therefore, co-extrusion and lamination can be sources of packaging reduction.

Plastic Packaging Products

Plastic Bag

A poly bag, also known as a pouch or a plastic bag, is manufactured out of flexible, thin, plastic film fabric. It is one of the common types of packaging and can carry a wide range of products including food items, flowers, waste, chemicals, magazines, and so on.

Poly bags are durable yet lightweight, reusable and flexible. Since poly bags are structurally simple to make, it can be fully customized in design, style & sizes but still remain cost-effective. Plastic recycling is also possible with poly bags, depending on the construction. Most of the poly bags are made with security features, tape attachments, hanging holes, and carrying handles to make sure the products are well secured and visually appealing to the customer.

Plastic Boxes

Plastic is used in a wide range of products, from spaceships to paper clips. A number of traditional materials, such as wood, leather, glass, ceramic, and so on, have already been replaced by plastic.

Plastic box packaging has many advantages in which they can be recycled, and generally they are much more durable than paperboard boxes. Airtight plastic packaging containers can help to preserve the quality of food and eliminate any contamination issues. Plastic packaging also does not break easily and can be stored with food under extreme conditions.

Another reason why plastic is a popular choice for packing material is because of its ability to showcase the product at any angle without necessarily opening the packaging. It is also flexible, lightweight and can be applied with films or coating to enhance packaging appearance.

Contrary to popular belief, plastic is in fact recyclable, in the sense that it takes less energy to produce new plastic, compared to glass, and other materials. Best of all, it is very cost effective.

Boil-in-bag

Boil bags or boil in bags as they are called are mainly vacuum sealed food bags that are used for preserving food items longer in a freezer. However, these bags are also extensively used in different parts of the world by many chefs for the purpose of cooking. The current models of boil bags are ideal for both cooking as well as storage. A wide range of items can be cooked in boil bags, including rice, ham, fish, salmon, beef, veal shank, zucchini, artichokes and other delicacies. They can be used for cooking when you go for weekend camping trips and they are also ideal for large scale commercial kitchen cooking. Easy to use, flexible and affordable, they can be used anywhere for the best results.

One of the best advantages of using boil bags for cooking and keeping food is that they completely eradicate the need for having additional cooking pans or storage containers. All one needs to have is hot boiling water and these bags can be used to make the food preparation that one wants. These bags are made out of industry approved materials that do not contain BPA or dioxin. They also come with special vacuum seals that prevent moisture and air from entering the bags and ruining the food. Many a times, keeping food in the freezer actually ruins the taste and flavor of the food even though it protects them from becoming stale. However, using boil bags perfectly preserve the flavors and taste of the food items while keeping them in perfect condition.

As leading restaurants as well as food and beverage stores are now acknowledging the benefits of using boil bags, more and more people are using them for their business and personal lives. Most companies manufacturing these bags offer them at wholesale prices, which mean that they are extremely cheap and affordable. Getting them in bulk also allows users to cook and store their food items in a more efficient and organized manner. Another distinct advantage that these boil bags have compared to conventional plastic storage containers or plastic bags is that they prevent the food from being exposed to oxygen and moisture. Oxygen and moisture, when combined with food can frequently give rise to the growth of mold and bacteria. On the other hand, making use of boil in bags offers a highly effective way to prevent the moisture and oxygen from damaging the food. The high level of safety standards followed by the makers of these bags has led them FDA and USDA approvals for widespread commercial usage.

The boil bags that are currently available in the market are not just suitable for home cooking; they can also be used for preparing gourmet meals in restaurants and also for catering purpose. These bags make use of sous vide cooking, which is a special type of cooking ideal for chicken and steak items while preserving their tender and juicy flavors. This enhanced flexibility with cooking is what makes professional chefs recommend boil bags for a large variety of preparations.

Edible and Bio based Packaging Materials

Biobased food packaging materials are derived from renewable sources and are potentially biodegradable that is composting (which is a technique for waste management). Biobased packaging materials include both edible coatings and edible films along with primary and secondary packaging materials. At the turn of the last century most non-fuel industrial products; dyes, inks, paint, medicines, chemicals, clothing, synthetic fibers and plastics were made from biobased resources. During the last years, the leading world research teams have been working on developing new biodegradable and edible packaging based on renewable biological sources, the so called "regulated life cycle materials". By the 1970s petroleum derived materials had, to a large extent, replaced those materials derived from natural resources. Recent developments are raising the prospects that naturally derived resources again will be a major contributor to the production of industrial products. Biobased /green polymers in food packaging are the wave of the future. The Scientific challenge is to find such applications and thus to create the demand for large scale production of biopolymers/ biomaterials that would help in attaining the sustainable development of green materials in contrast to petroleum.

References

- The-Environmental-Impacts-of-Packaging-229796182: researchgate.net, Retrieved 22 May 2018

- Molded-pulp-packaging-pulp-products: advancedpaper.com, Retrieved 29 April 2018

- Function-advantage-egg-trays: eggtraymachines.net, Retrieved 31March 2018

- Excellent-benefits-of-using-aluminium-foils-for-packaging-055311: boldsky.com, Retrieved 11 April 2018

- Boil-bags-indispensable-to-cooking-storing-food: universalplastic.com, Retrieved 21 June 2018

- Biobased-Packaging-Materials-for-the-Food-Industry-258769065: researchgate.net, Retrieved 28 June 2018

Types of Food Packaging Machinery

There are diverse machineries that are used for food packaging. Some of these are multihead weigher, sealing machines, case sealer, cartoning machine, vacuum-packaging machines, wrapping machines, etc. which have been carefully examined in this chapter.

Depending on the type of food being packed, packing comes in various types. To pack these food materials, various food packaging machines are used. The packing styles also change depending on the storage life of the product.

Food that are high perishable like fresh processed meats and frozen items are best when vacuum packed since it can tremendously extend its storage life. There is a separate type of food packaging machine or food packing equipment used to perform vacuum packaging of the products.

Various Types of Food Packaging Machines

Food Vacuum Packaging Machine

It is one of the most efficient packaging machine to pack foods because it avoids air making food remain fresh. As aerobic microorganisms are responsible in swift deterioration of foods, they hardly thrive or are immobilized under this condition.

Food vacuum packaging machine helps to extend storage life of food products thereby making the product well suited for sale on the freezer or cold display storage units of several retail stores.

Biscuit Packaging Machine

Biscuit packaging machine is another type of food packaging equipment. It is usually fitted with electronic digital temperature controller to maintain high precision in achieving the desired temperature during food packing process.

It helps to bring optimum freshness of the food. The most interesting aspect of this machine is that packaged products are closely monitored with its automatic feed counter that shows the quantity of items placed packed by machine. This makes it easier for the food manufacturing companies to monitor daily factory output.

Bundling Food Packaging Equipment

Bundling Food packaging Equipment is quite common and is widely used by many food suppliers. It is capable of storing huge quantity of foods before they are banded or wrapped together as a single bundle.

It is also called as the banding machine. It can also be used for packing small items such as stick candies or individually packed hot-dogs that need to be bundled together for economic purposes.

Bagging Machine

It is popular in several China food processing factories. Foods in this case are packed in bags, sacks and pouches. This bagging machine is common to pack cereals and powdered foods such as milk powder and sugar.

Closing Machines

These closing machines are similarly common in many food factories. This equipment is used to tie metal wires to enclose the food bag or pouch.

Capping Machines

Capping machines are popular among food suppliers of food syrups and drinks. This equipment is not used solely to pack food items but it is usually used in conjunction with other food packaging equipment.

The major function of this equipment is to close bottled food items by placing air-tight caps. This is common in soda-manufacturing companies.

Accumulation Machinery

Accumulation machinery is used along with the capping machine. This machine allows proper alignment of bottles for systematic and organized filling of foods. It is used in soda companies and bottled-water companies.

There are various types of food packaging machines. It is important to tailor-fit the selection when yo buy one of these machineries according to the type of food that is being packaged to ensure optimum quaslity products with fully extended storage life.

Multihead Weigher

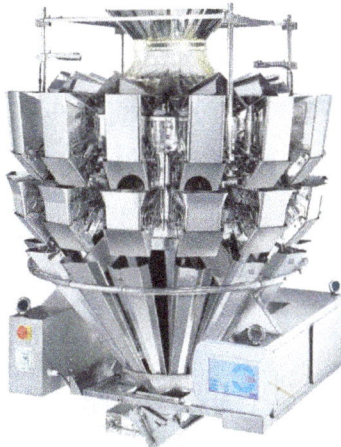

Multihead weighers offer food manufacturers a vital edge as they strive to meet the requirements of their customers. With supermarkets and the food service sector insisting on ever more exacting specifications, it is crucial that food manufacturers maintain consistent quality on their production lines.

The vast majority of food products are sold by weight, so measuring out precise quantities, quickly, time after time, with as little waste or giveaway as possible, is vital, and this is exactly what multihead weighers are designed to do.

The first multihead weigher was developed by Ishida in 1972 and although there have been many innovations and enhancements since then, the 'combination method' of how they work has remained the same.

The weighers operate by having a number of hoppers arranged in columns, and each column is fitted with a weighing head. A portion of the product to be weighed is fed into each weigh hopper and the machine's computer evaluates the optimum combination of hoppers to make up the target weight. These are then discharged together before the hoppers are recharged from above – ready to make the next weighing combination.

This method produces a high level of accuracy when weighing single products, and is used across a wide range of applications from snacks and confectionery to grated cheese, salads, fresh meat and poultry.

Pin-point Accuracy

Using larger machines with more weighing heads allows the same principles to be used to provide pin-point accuracy for mix-applications. Each ingredient within a mix can be assigned to a specific section of weighing heads within the machine. They will measure out the required amount of each ingredient and then release them, in most cases simultaneously, via the multihead weigher's discharge chute into the pack.

Whether it is a pack of mixed sweets, mixed cereals, nut mix, or frozen vegetables, Ishida has a range of specially designed mix weighing solutions with 16, 20, 24, 28 and 32 weighing heads, which enable up to eight different components to be mix-weighed so consistent amounts of each ingredient are included in each portion of the mix.

Such accuracy means brand consistency is maintained and legal requirements when stating percentages of ingredients in a mix can be met. Manufacturers can also control the ratios of products in each portion to manage the quantity of the most expensive ingredients in any mix and to maximize profits.

Another advantage of multihead weighers, particularly when used in mixing applications, is the small factory footprint of the machines. When compared with a production line of individual machines that would be required to do the same job they take up far less valuable floor space. Having one machine instead of several also means maintenance and labour costs are minimised as only one person is needed to monitor the multihead weigher, while several would be needed to oversee a series of machines on the line.

Having all products processed by one weighing machine rather than a line of machines dropping product in sequence also allows products to be mixed effectively – avoiding layering of components if necessary.

A vast range of products with many different characteristics can be handled in mixed weighing. Products could vary from free-flowing components to sticky or fragile components all within same mix. To ensure they are all processed by the weighers at the highest possible speed and efficiency levels, Ishida has developed a wide range of filling systems tailored to each product component characteristic and/or pack type. By adopting a variety of vibration control settings, different contact parts designs, gentle handling options as well as specialist materials and angles, Ishida ensures even the most challenging products pass through the machine at the optimum rate. Where product runs vary, Ishida can also offer optional change contact parts on each section to adapt quickly to many different production challenge requirements.

Sealing Machines

Sealing machines close and seal an individual package or provide a long continuous horizontal or vertical seal. There are many different types of sealers. Sealing machinery that combines form, fill, and seal functions is also available. Some sealing machines transport the plastic film horizontally, while others transport vertically.

- Seal-only equipment wraps or secures products, but does not form packages or fill them.

- Manual sealing machines aid only in the setting or holding of products.

- Semi-automatic sealers help with both packaging and placement, allowing a single operator to perform several activities with greater speed and accuracy.

- Fully automatic sealing equipment requires limited operator intervention. Typically, operators need only replenish packaging components by loading supply hoppers or removing completed cartons. One of the main differences between a semi-automatic and fully automatic sealing machine is that fully automatic sealers close all of the flaps, including the leading. trailing, and side flaps before taping, while semi-automatic sealers tape only the top and bottom.

Case Sealer

Carton sealing machines, commonly known as case sealers, are packaging machines that fold and seal the top lids of the packed cartons. These machines offer a reliable and efficient way to seal cases after the packaging process.

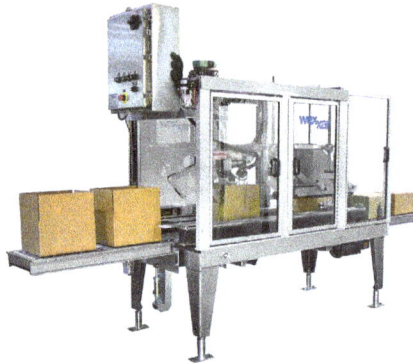

An automatic sealing machine is a safe way to fulfil the packaging needs. Removing the need for glue guns and sharp blades means the staff will be working in a far safer environment.

As these machines are used at the tail end of the production line, their importance cannot be understated. Without having a high-quality sealing machine in place, you are running the risk of holding up the whole plant. Therefore, you really need a piece of kit you can rely on if you want to avoid costly bottlenecks and get the product out of the door in a timely fashion.

There's also a money saving element, too. Opting for an automatic box sealing machine can save you cash in labour costs over the long term, so these packaging sealers can really be regarded as an investment in the business rather than just another cost.

Heat Sealer

After trays are filled, they need to be sealed. Heat sealers apply plastic film to multiple conveyor lines of trays by using a web of film to cover the trays. The film is cut and adhered to the trays by the heating element, which is lowered to meet the trays when they are in position. The leftover film on the web then continues on and is spooled onto a waste reel.

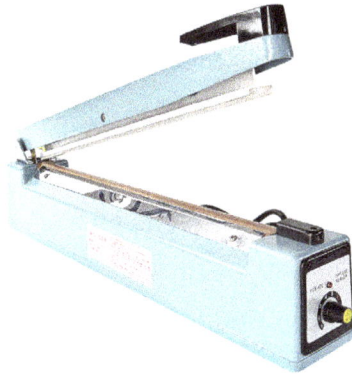

These types of machines present multiple opportunities to keep production efficient and throughput high. The roll diameter of the film must be monitored to make sure that product does not run out or jam and result in costly downtime. Accurate roll diameter monitoring allows timely replacement of rolls before they run out, but not too soon so that excess material is not wasted.

Check Weighing Machines

A checkweigher is a system that weighs items as they pass through a production line, classifies the items by preset weight zones, and ejects or sorts the items based on their classification. Checkweighers weigh 100% of the items on a production line.

Typically, an infeed section, scale section, discharge section, rejector or line divider, and computerized control comprise the physical checkweighing system. Checkweighers and their components vary greatly according to how they are used, the items being weighed, and the environment surrounding them.

Simply stated, a checkweigher weighs, classifies, and segregates items by weight.

Many possible uses for a checkweigher include:

- Check for under and/or overweight filled packages.

- Insure compliance with net contents laws for prepackaged goods.

- Check for missing components in a package including labels, instructions, lids, coupons, or products.

- Verify count by weight by checking for a missing carton, bottle, bag, or can in a case.

- Check package mixes against weight limits to keep the solid to liquid ratio within established standards.

- Reduce product giveaway by using checkweigher totals to determine filler adjustments.

- Classify products into weight grades.

- Insure product compliance with customer, association, or agency specifications.

- Weigh before and after a process to check process performance.

- Fulfill USDA or FDA reporting standards.

- Measure and report production line efficiency.

Filling Machines

Filling machines are equipment used for packaging of various products, mainly food and beverages. Depending on the product, the container to be filled can either be a bottle or bag. These machines are usually found in manufacturing industry to promote quality and efficiency on the manufacturing process. Products such as sugar, cereals and fresh milk are available in certain quantities. Filling machines are set to fill up the cartons, plastic bags or bottles with the exact amount of product designated to each of them.

Being a usual part of an assembly line system, filling machine is a vital component of a bottling process. This type of filling machine is present on facilities that produce liquid based products such as beverages, chemicals and household cleaners. The bottling process includes filling, labeling and packing equipment. These equipment are incorporated with a bottle conveying system which is responsible in transporting bottles to go through the different processes. The bottle conveyor will make sure that the bottles are held securely to keep them aligned with the filling machine.

There are several types of *filling machine* used in various packaging industry. The types which are commonly utilized in the production of goods are liquid filling machine, paste filling machine, powder filling machine and granular filling machine. Liquid filling machine is applied in the production of the liquid-based products such as carbonated drink, perfume, alcoholic beverages, shampoo, oil and so on. It is convenient to use and easy to hold. It is suitable in filling sticky paste products as it can also feed hopper.

Paste filling machine is the most sough out type of filling machine in the market today. The reason behind this is because of its perfect appearance. It is also exquisitely made and performs in a very unique way. Applied in the production of cosmetics, paste filling machines are utilized as well in the field of sauces, ointment, honey and other paste products.

From the name itself, powder filling machines are used in the manufacture of any powdery materials such as powdered milk, flour, chemical agent, pesticide and a lot more. Similar to paste filling machine, this type is designed exquisitely. It follows the augur filling principle which marks the beginning of innovation in how filling machines work. Auger filling machines have cone shaped hopper that holds the dry mixes or the powdery materials. They will be poured inside a pouch made of paper or poly forming a ring with the use of an auger screw controlled by the agitator. The pouch is then sealed through a series of dies and heaters.

Granular filling machinery is similar to powder filling machine yet considered to be more efficient to use in granular fill processes. It is applied in packaging of all kinds of granules such as medicine, pesticide, sugar, coffee, tea and others.

Barcode Printer

A barcode printer is a printer designed to produce barcode labels which can be attached to other objects.

A dynamic barcode system has multiple applications for today's foodservice companies:

- Tracking the distribution of food shipments as they move from warehouse to final destination.

- Recording expiration dates on individual shipments.

- Tracing contamination that originates off site.

- Reducing invoice discrepancies.

- Eliminating waste due to over-purchase of products.

Barcode Reader

A barcode reader, also called a price scanner or point-of-sale (POS) scanner, is a hand-held or stationary input device used to capture and read information contained in a bar code. A barcode reader consists of a scanner , a decoder (either built-in or external), and a cableused to connect the reader with a computer. Because a barcode reader merely captures and translates the barcode into numbers and/or letters, the data must be sent to a computer so that a software application can make sense of the data. Barcode scanners can be connected to a computer through a serial port , keyboard port , or an interface device called a wedge . A barcode reader works by directing a beam of light across the bar code and measuring the amount of light that is reflected back. (The dark bars on a barcode reflect less light than the white spaces between them.) The scanner converts the light energy into electrical energy, which is then converted into data by the decoder and forwarded to a computer.

There are five basic kinds of barcode readers -- pen wands, slot scanners, Charge-Couple Device (CCD) scanners, image scanners, and laser scanners.

- A pen wand is the simplest barcode reader. It contains no moving parts and is known for its durability and low cost. A pen wand can present a challenge to the user, however, because it has to remain in direct contact with the bar code, must be held at a certain angle, and has to be moved over the bar code at a certain speed.

- A slot scanner remains stationary and the item with the bar code on it is pulled by hand through the slot. Slot scanners are typically used to scan bar codes on identification cards.

- A CCD scanner has a better read-range than the pen wand and is often used in retail sales. Typically, a CCD scanner has a "gun" type interface and has to be held no more than one inch from the bar code. Each time the bar code is scanned, several readings are taken to reduce the possibility of errors. A disadvantage of the CCD scanner is that it cannot read a bar code that is wider than its input face.

- An image scanner, also called a camera reader, uses a small video camera to capture an image of the bar code and then uses sophisticated digital image processing techniques to decode the bar code. It can read a bar code from about 3 to 9 inches away and generally costs less than a laser scanner.

- A laser scanner, either hand-held or stationary, does not have to be close to the bar code in order to do its job. It uses a system of mirrors and lenses to allow the scanner to read the bar code regardless of orientation, and can easily read a bar code up to 24 inches away. To reduce the possibility of errors, a laser scanning may perform up to 500 scans per second. Specialized long-range laser scanners are capable of reading a bar code up to 30 feet away.

Label Printer Applicator

Label Printer Applicators are used for applying human readable and barcode information to many different products, generally cartons or pallets.

The major benefits of Label printer applicator include their ability to:

- Eliminate the expense and inconvenience of preprinted labels

- Eliminate label obsolescence

- Reduce labor costs

- Provide faster, more accurate labeling.

Can Seamer

The fully automatic can seaming machine was designed and built for seaming round tin, aluminium or composite cans. It works on the principle of a rotating can.

Technical advantages:

- Sophisticated design made of stainless material

- Suitable for wet and dry undertakings

- Hygienic and easy to clean

- Flexible working by automatic height adjustment

- Constant working height

- Fully automatic can and lid feed

- Standstill clamping and unclamping of the can

- Large, maintenance-free seaming rollers.

Cartoning Machine

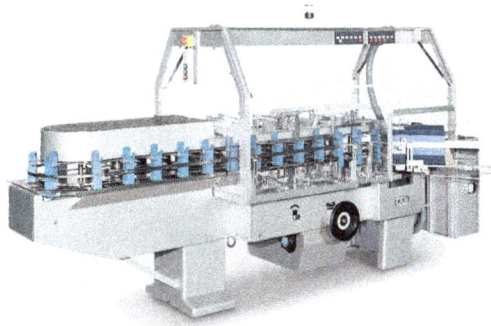

A cartoning machine is an integration of air, electrical, light and mechanical high-tech equipment configured to a computer to control, manage and regulate its operation.

Fundamentally, it works by forming cartons:

- It shapes cartons out of blanks to stand up-straight

- Close or/and erect

- Introduce a product

- Folds

- Side-seams and finish by sealing.

Below are a few reasons to why you need a cartoning equipment in a production environment.

- Product protection and ease of transport

  Carton provides protection to products filled in them. Since cartons assume regular shapes, makes it easier for bulking packaging for lorries, containers and other vessels easing transportation.

  For instance, bottles need a proper carriage to accommodate its unique shape. However, with an appropriate cartoning machine such as ZH-100P Cartoning Machine for Bottle&Vial we can achieve a better way to pack bottles.

- Save time

  High speed production lines are susceptible to time wastage during packaging. However, with continuous mode cartoning machine high speed is achievable and guarantees high throughput altogether.

  In case you have gone for a fully-automatic cartoning machine that means there will be reduced labor-input during the production process. And since machine does not suffer fatigue unlike man, production can run for longer hours without unnecessary breaks, ultimately, high throughput.

- Tamper proof

  Apart from protecting the products from adulteration and interference, carton packaging can also act as tamper proof. Carton packaging done by aid of human and a machine show stark differences from ones done with a Cartoner.

  A cartoning equipment are designed to perform each task in trig and typically same manner, which can be vividly observed. The unique manner in which the cartoning equipment seal the cartons can aid in determining whether there is tampering with any product. This reason also builds consumer confidence in your brand.

- Handle a vast array of commodities

  If you are concerned about multi-packaging, then you should cast away your aspersions as some semi-automatic cartoning equipment can multitask. For instance, products such as tooth paste and brush can be simultaneously packaged into single carton.

Vacuum-packaging Machines

A vacuum machine for food packaging is essential in different situations: at home, in restaurants, grocery stores and delicatessens, but also in large companies and even food processing laboratories.

- Vacuum packaging extends shelf life, ensures quality, prevents products from drying out, protects it from outside influences and improves hygienic handling.

- Vacuum packing is a very cost effective procedure for preserving perishable food products.

- Very low packing material costs, short process times and a relatively low investment for long-lasting equipment makes vacuum packing by far the most economical and most efficient way of packaging perishable food products.

Wrapping Machines

Wrapping Machines wrap a flexible packaging material, (e.g. paper, aluminium, plastic film), around a product or group of products.

A major application is in the field of shrink wrapping where heat is applied by various means to a thermoplastic material already loosely wrapped around the product or group of products, which then shrinks around them to form a tight wrap. This method is often used for transit wrapping and protective packaging of larger item such as doors or even bricks on a pallet.

Because wrapping is so versatile it is used in many sectors, however, it is most common in food, bakery and confectionery for single items which can range from confectionery (count line), bars and cakes through to cheese and sausages.

References

- Types-of-food-packaging-machines: packaging-labelling.com, Retrieved 28 June 2018

- Promotional-Features, Multihead-weighers-add-consistency-to-the-mix-2: foodmanufacture.co.uk, Retrieved 18 July 2018

- Types-of-filling-machines: thegreenbook.com, Retrieved 18 May 2018

- Barcode-reader-POS-scanner-bar-code-reader-price-scanner: techtarget.com, Retrieved 21 May 2018

- Label-printer-applicators: weberpackaging.com, Retrieved 31 March 2018

- Cartoning-machine: saintytec.com, Retrieved 10 April 2018

- Vacuum-machine-for-food-packaging-machine-for-all-needs: minipack-torre.it, Retrieved 22 April 2018

5

Controlled and Modified Atmosphere Packaging

To improve the shelf life of food products, the composition of the internal atmosphere of a package is modified. There may be a reduction in oxygen or replacement of it with different gases. The concentration of gases such as carbon dioxide and nitrogen may also be controlled. This chapter discusses in elaborate details the crucial aspects of controlled and modified atmospheric packaging, such as gas flush, foodstuffs in MAP and gas technology for modified atmosphere packaging.

Controlled Atmosphere Packaging

Controlled atmosphere storage is the storage in which atmosphere of oxygen, carbon dioxide and nitrogen (and sometimes other gases) are controlled by external control systems.

Another term sometimes used in 'intelligent' packaging which may include monitoring the package and possibly making adjustment based on this information.

Controlled atmosphere (CA) storage has been shown to be a technology that can contribute to these consumer requirements in that in certain circumstances, with certain varieties of crop and appropriate treatment, the marketable life can be greatly extended.

Advances in controlled atmosphere storage technology include faster establishment of desired atmosphere, less fluctuation in O_2 and CO_2 levels, ability to change atmospheric composition as needed during storage, and ability to scrub ethylene from the storage environment.

Commercial use of controlled atmosphere storage is greatest in apples and pears world-wide; less on kiwifruits, avocados, persimmons and pomegranates, nuts and dried fruits and vegetables.

Controlled atmosphere packages allow reaching markets that are geographically far from the point of packaging for sensitive materials. Indeed, the sophistication of such packages will increase costs and the trade-off with alternative technologies such as refrigeration will likely be determined by balancing convenience and economics.

Modified Atmosphere Packaging

Modified Atmosphere Packaging (MAP) is well established in the food industry and continues to gain in importance. MAP means, simply put, that the natural ambient air in the package is replaced by a gas or gas mixture, often nitrogen and carbon dioxide. This packaging under a protective atmosphere preserves the quality of fresh produce over a longer period of time, prolongs shelf-life, and gives food producers access to a geographically larger market for perishable products. This is suitable for meat and sausage products, dairy products, bread, fruits and vegetables, fish or convenience products.

Modified atmospheres are not only used in packaging. They can feature as part of the production process, e.g. in the case of minced meat, or of storage and transport, for example of fruit and vegetables in halls or containers.

The standards required by Modified Atmosphere Packaging are comparatively high, and have to be controlled and monitored to ensure safety. Therefore, food manufacturers rely on modern MAP gas technology and various levels of quality assurance for maximum process safety.

Benefits of Modified Atmospheres

- Longer shelf life / higher quality

Food packaged under a protective atmosphere spoils much slower. Combined with continuous cooling, Modified Atmosphere Packaging can significantly extend the freshness and shelf life. This effect varies depending on the product type. However, a doubling of the shelf life is usually possible. Normally, MAP products keep a high quality over a longer period of time and arrive at the consumer in the best possible condition.

- Less waste

Longer durability is often associated with fewer problems during long distance shipment, and longer shelf life. As a result, waste disposal due to spoiled food can often be reduced.

- More sales opportunities

Because of the longer shelf life, Modified Atmosphere Packaging typically opens up new geographic markets to manufacturers. Particularly with perishable goods, longer shipment distances can be achieved. A global market can become a reality.

- Fewer preservatives

Packaging under a protective atmosphere extends the shelf life of food, meaning in many cases that the use of preservatives can be reduced or even completely eliminated. Consumers get products that do not contain artificial additives.

- Appealing package design

Next to functional aspects, the design of the packaging plays a significant role in the competition for consumers. The look-and-feel and the quality impression influence the purchasing behavior. Modified Atmosphere Packaging is very well suited for the most appealing packaging design and presentation of the food product.

Limitations of Modified Atmospheres

- Comparatively high complexity

The modified atmosphere packaging process involves comparatively high requirements. Possible failures: incorrect gas composition or leaks due to faulty temperature or pressure distribution, contaminated or worn tools, seal contamination or defective material. However, with modern MAP technology and comprehensive quality assurance, the risks can be mastered.

- Relatively high cost

In addition to high-quality films, the consumption of gas and the personnel costs for quality control are particularly costly. However, these costs can be minimized with efficient use of resources.

- Influence on product quality

Unlike using preservatives, in most cases, the protective gases are not absorbed by the food and thus do not alter the nature or taste of the product. But there are exceptions to this rule. For example, an excessively high concentration of CO_2 can be absorbed by the food and make it sour. However, these effects can be avoided with adapted gas mixtures. The influence of very high oxygen concentration on the quality of meat is controversial. Modified atmospheres are supposed to make the meat more chewy.

Factors Influencing the Shelf Life of Food and the Influence of Modified Atmospheres

From the time that fruits and vegetables are harvested or animals are slaughtered, the spoilage process begins. This process is often accelerated the more the products are processed, such as cut fruit or minced meat. How long foods are durable, which means suitable for consumption, is very different and depends on various factors, e.g. the content of water and salt, pH value, hygiene conditions during production, storage conditions such as temperature or humidity, packaging. Depending on the characteristics and combinations of these factors, food products are differently sensitive to microbial or chemical/biochemical spoilage.

Chemical and Biochemical Spoilage

Directly after harvesting of plant or slaughter of animal material, chemical processes begin to change the structure or quality. Sometimes this is useful, e.g. dry-aging of meat, which can be seen as a maturation to improve quality. In principle, however, the quality of organic material decreases. For example, the oxidation of fats quickly leads to a rancidity of the product.

Microbial Spoilage

Microorganisms are a major threat to the shelf life and quality of food. On the one hand, they influence color and smell, but they can also lead to health hazards and make the products uneatable. The source of the microorganisms is either the food itself or an impurity that cannot be completely excluded in the production and packaging process.

The changes due to chemical / biochemical and microbial spoilage can be significantly slowed by MAP techniques together with cooling. Various gases and mixtures with different properties are used to slow the process of spoilage as much as possible.

Quality Control of Modified Atmosphere Packaging

Modified Atmosphere Packaging makes comparatively high demands on the packaging process, especially on the sealing process. Many sources of error can lead to leaks, usually micro-leaks. Right from the point of mixing the gases and introducing them into the package, maximum care is required. A faulty mixture or leaking packaging can have serious effects - from the loss of nutrients, taste, color or structure to a bad smell or infestation with microorganisms. Depending on the product, health risks to the consumer cannot be completely eliminated.

Modified Atmosphere Packaging therefore requires modern high quality equipment and uncompromising standards of hygiene. But even when using best available technology, failures can't be completely avoided. So comprehensive quality assurance activities are essential. These can begin during the packaging process, using inline gas analysis, which constantly monitors the composition of the modified atmosphere. After packaging, the packages must be tested for the correct gas mixture and for leaks. Only with this level of rigor can it be ensured that the full benefit of Modified Atmosphere Packaging is achieved and that the customer receives a top quality product.

Gas Flush

Gas flush consists of an inert gas such as nitrogen, carbon dioxide, or exotic gases such as argon or helium which is injected and frequently removed multiple times to eliminate oxygen from the package. This technique is called MAP (Modified Atmosphere Packaging). Most common applications for MAP include coffee, snack foods, pre-baked products, meat and poultry, as well as other more sophisticated applications.

When modifying the atmosphere inside of a package, the amount of oxygen can typically be reduced to 3% or less. Inert gases used for MAP are typically denser than oxygen. As such, the oxygen inside the package is forced out of the package. This results in extended product shelf life, product integrity, protection against discoloration, and for products like chips, a cushion-like buffer against damage (this is commonly referred to as a "pillow pack").

Many gas flush applications require sophisticated gas mixtures. Case ready meat applications are a good example of a sophisticated gas mixture which requires a nitrogen, carbon dioxide and carbon monoxide mixture (Tri-Gas). Nitrogen is an inert gas that functions to fill the headspace in the package. Carbon dioxide is added for its anti-microbial properties and carbon monoxide stabilizes the typical red or pink color of air-exposed meat.

Some companies have used dry nitrogen to reduce the size of a desiccant packet in high volume semi-conductor packaging applications spread over 300,000 - 400,000 packages. The cost reduction for the total package or sorbent system easily results in savings that could provide the capital for the vacuum sealing equipment.

Common MAP Gases

Carbon Dioxide (CO_2)

Carbon dioxide inhibits the growth of most aerobic bacteria and molds. Generally speak-ing, the higher the level of CO_2 in the package, the longer the achievable shelf life. However, CO_2 is readily absorbed by fats and water - therefore, most foods will absorb CO_2. Excess levels of CO_2 in MAP can cause flavor tainting, drip loss and pack collapse. It is important, therefore, that a balance is struck between the commercially desirable shelf life of a product and the degree to which any negative effects can be tolerated. When CO_2 is required to control bacterial and mold growth, a minimum of 20% is recommended.

Nitrogen (N_2)

Nitrogen is an inert gas and is used to exclude air and, in particular, oxygen. It is also used as a balance gas (filler gas) to make up the difference in a gas mixture, to prevent the collapse of packs containing high-moisture and fat-containing foods, caused by the tendency of these foods to absorb carbon dioxide from the atmosphere. For modified atmosphere packaging of dried snack products 100% nitrogen is used to prevent oxidative rancidity.

Oxygen (O_2)

Oxygen causes oxidative deterioration of foods and is required for the growth of aerobic micro-organisms.

Generally, oxygen should be excluded but there are often good reasons for it to be present in controlled quantities including:

- Maintain fresh, natural color (in red meats for example)

- To maintain respiration (in fruit and vegetables)

- To inhibit the growth of aerobic organisms (in some types of fish and in vegetables).

Argon

Argon has the same properties as nitrogen. It is a chemically inert, tasteless, odorless gas that is heavier than nitrogen and does not affect micro-organisms to any greater degree. It is claimed to inhibit enzymic activities, microbial growth and degradative chemical reactions. Hence it can be used in a controlled atmosphere to replace nitrogen in most applications. Its solubility (twice that of nitrogen) and certain molecular characteristics give it special properties for use with vegetables. Under certain conditions, it slows down metabolic reactions and reduces respiration.

Carbon Monoxide

Carbon monoxide is a toxic, colorless, odorless, flammable gas. It is stable at up to 400°C with respect to decomposition into carbon and oxygen.

Results have shown that the use of carbon monoxide (CO) in MAP with high levels of CO_2 has resulted in increased shelf-life together with retention of the bright red color of meat cuts. It is also claimed that carbon monoxide can effectively reduce or inhibit different spoilage and pathogenic bacteria.

Foodstuffs in MAP

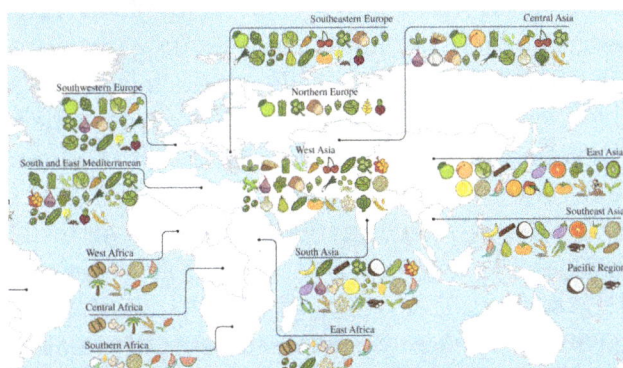

Modified Atmosphere Packaging is suitable for a wide range of food product. While traditionally mainly dairy products, meat products or bread were packaged under protective atmosphere, now MAP is more and more used for other foods like fish, coffee, fruit or vegetables. In addition, Modified Atmosphere Packaging is driven by the growing popularity of ready-made meals and convenience products.

Meat and Sausage Products

Meat and sausage products, especially raw meat, are very prone to spoiling due to microbial growth, on account of their high moisture and nutrient content. No matter whether beef, pork or poultry – spoilage begins from the moment of slaughter and especially after butchering. Besides high hygiene standards and permanent cooling, modified atmospheres can significantly extend the shelf life of meat and sausage products. CO_2 is the most important among the protective gases. At concentrations above 20 %, CO_2 can considerably reduce microbial growth. In the case of red meat, there is also the risk of oxidation of the red colour pigments. The meat will lose its red colour, becoming grey and unappetising in appearance. This oxidation is especially prominent with beef. A high oxygen content in protective gas packaging can prevent oxidation. A low carbon monoxide content (approx. 0,5 %) can also help to retain the red colour of meat. However, the use of this gas is not allowed in the EU, for example. Poultry is especially sensitive to rapid spoilage and is therefore subject to higher requirements for permanent cooling. Here too, a modified atmosphere with CO_2 content will extend the shelf life. A high oxygen content is also used for poultry without skin so as to retain the colour of the meat. The CO_2 can partly be absorbed by the foods. To prevent the packaging from collapsing, nitrogen is used as a supporting gas.

Sausage and meat products, e.g. marinated or smoked meat pieces, react very differently depending on the preparation. Longer shelf-lives can be afforded by the use protective gases right from the start. The CO_2 content should not be too high with these products, in order to prevent a sour taste.

Fish and Seafood Products

Fish and seafood are some of the most sensitive foods. They are at risk of rapidly declining in quality and spoiling even shortly after the catch. The reason for this lies in the neutral pH value as an ideal precondition for microorganisms as well as special enzymes that negatively affect taste and odour. Fish, which is rich in fatty acids, also becomes rancid quickly. The most important element for a longer shelf life is cooling close to 0° Celsius. Modified atmospheres with minimum 20 % CO_2 also retard the growth of bacteria. CO_2 components around 50 % are frequently used. Higher CO_2 concentrations can lead to undesirable side effects such as liquid loss or a sour taste. In the case of low-fat fish and shellfish, O_2 is also used in the packaging. This prevents a fading or loss of the colour, while at the same time serving as a growth inhibitor for some types of bacteria. When dealing with shellfish and crustaceans, special attention should be paid to ensuring a CO_2 content that is not too high. This can be discerned most clearly by a sour taste, while these products absorb CO_2 the most, as a result of which the packaging can collapse. Nitrogen as an inert supporting gas prevents this effect.

Dairy Products

Cheese is predominantly spoiled by microbial growth or rancidness. A continuous cooling chain essentially extends the shelf life of products. With hard cheese, there is a risk of mould formation upon contact with oxygen. As a result, vacuum packaging was frequently used in the past, even though these are awkward to open and can leave unattractive marks behind on the product at the same time. CO_2 effectively prevents mould formation, but does not otherwise affect the maturation of the cheese. Soft cheese can quickly become rancid. This problem can also be tackled with CO_2 modified atmospheres. However, as soft cheese absorbs CO_2 to a significantly higher extent, there is a risk of the packaging collapsing. A correspondingly lower CO_2 content should therefore be chosen. In the case of milk products such as yoghurt or cream, there is a risk of the products absorbing too much CO_2 and becoming sour. A lower CO_2 content should therefore be chosen.

Milk powder, above all for use in baby food, is a highly sensitive product. It is especially important to ensure that oxygen is displaced from the packaging in order to extend the shelf life. In practice, packaging is carried out in pure nitrogen with as low a residual oxygen content as possible.

Bread and Cake

With bread, cake and biscuits, the shelf life is primarily affected by potential mould formation. A high standard of hygiene during production and packaging can significantly minimise this risk. Packaging involving a modified atmosphere with CO_2 and without oxygen largely prevents the products from becoming mouldy and extends the shelf life. To prevent the packaging from collapsing owing to CO_2 absorption by the products, nitrogen is used as a supporting gas in many cases.

Fruit and Vegetables

Modified atmospheres in packaging make it possible to offer consumers fresh and untreated products – in other words succulently fresh fruit and vegetables – with a long shelf life. At the same time, fruit and vegetables are subject to very special requirements in regard to the nature of the packaging and atmosphere. This is because – in contrast to other food – fruit and vegetables continue breathing after the harvest and consequently require an oxygen content in the packaging. Furthermore, the packaging film does not have to be fully tight. By taking the product's breathing and the permeability of the film, typically via micro-perforation, into account, the composition of carbon dioxide, nitrogen and low amounts of oxygen ideal for the product can be maintained. The term used here is an EMA (equilibrium modified atmosphere). The gas composition is individually adapted to the corresponding product.

Thorough cleaning along with hygienic processing are the fundamental preconditions for long-lasting freshness. Modified atmospheres, in conjunction with corresponding cooling, can be used to extend the shelf life of fresh produce, while achieving an attractive packaging design at the point of sale.

Pasta and Ready-made Meals

The nature and composition of fresh pasta and, in particular, readymade meals are very different. Above all, multi-component products such as ready-made pizzas or sandwiches contain many different foods with differing shelf lives and spoilage properties. In the majority of cases, modified atmospheres can significantly extend the shelf life without using oxygen. Mixtures of CO_2 and nitrogen are used here. The concentration of the gases is oriented to the content of the product. If, for example, there is a risk that large volumes of CO_2 will be absorbed by the product, the nitrogen content should be chosen higher to prevent the packaging from collapsing.

Snacks and Nuts

Snack products, for example potato crisps or peanuts, primarily involve problems associated with the fat content of the food. There is a risk of oxidation, whereby the products can quickly become rancid if the packaging is not optimal. To extend the shelf life, it is therefore important to minimise the contact with oxygen. Modified atmospheres with 100 % nitrogen are frequently used. In this way, a premature spoilage can be prevented, while these atmospheres also provide protection from mechanical damage to sensitive products, e. g. potato crisps in conventional packets.

Wine

Gases or gas mixtures are often used to protect wine in the different phases of its production process and to retain the quality of the product. They are mainly used to avoid contact with oxygen and prevent microbial deterioration. The tank headspace is replaced with an inert gas or a gas mixture, for example of CO_2, N_2 or Ar. The composition of the gases is chosen according to the type of wine.

Coffee

As a dried product, coffee is relatively insensitive to spoilage by microorganisms. However, the risk of the fatty acids it contains oxidising and making the product rancid is greater. To prevent this, oxygen is excluded from coffee packaging. Instead, a modified atmosphere comprising pure nitrogen is frequently used in coffee sachets or capsules.

Examples of Gas Mixture Compositions

Product	O_2	Co_2	N_2
Raw offal	80	20	0
Raw poultry with skin	0	30	70
Raw poultry without skin	70	20-30	0-10
Cooked meat and sausage products	0	20-30	70-80
Raw low-fat fish	20-30	40-60	20-40
Raw high-fat fish	0	40	60
Cooked / smoked fish	0	30-60	40-70

Shellfish and crustaceans	30	40	30
Hard cheese	0	30-100	0-70
Soft cheese	0	10-40	60-90
Sliced cheese	0	30-40	60-70
Cream cheese	0	100	0
Yoghurt	0	0-30	70-100
Milk powder	0	0-20	80-100
Crispy breads	0	50-100	0-50
Cakes, buicuits	0	50	50
Fruit / Vegetables, fresh	3-10	3-10	80-90
Vegetables, cooked	0	30	70
Ready-made meals	0	30-60	40-70
Pasta/Pizza	0	30-60	40-70
Sandwiches	0	30	70
Snacks/Crisps/Peanuts	0	0	100
Wine, white / Rosé	0	20	80
Wine, red	0	0	100
Coffee	0	0	100

Gas Technology for Modified Atmosphere Packaging

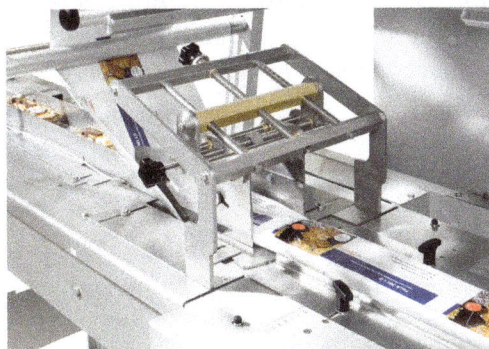

Packaging Machines

There is no one special packaging machine for Modified Atmosphere Packaging. Various types of machines from several suppliers do the job.

Hand vacuum chamber machines are the most simple type of MAP machines. They are operated manually and are suitable especially for smaller companies. Pre-formed bags are put into the chamber and filled with the product. After closing the chamber, the machine creates a vacuum and replaces the air with the modified atmosphere before the packaging is finally sealed.

For larger packaging volumes normally automatic packaging lines are used. So called thermoform-fill-seal machines are using packaging film from a roll. The film is heated inside the machine and formed onto trays which get filled with food product. The next steps are similar to the hand chamber machine but are done automatically. In a vacuum chamber the air is replaced by a gas mixture. The trays are then sealed. Tray-sealer machines function in a similar way. Main difference: The trays are not made inside the machine but are pre-formed and just sealed with a film.

Form-fill-seal or flow-pack machines are a further type of machine. Horizontal or vertical machines are available. These machines form a tube from a film and place the product inside. The air inside the tube is replaced by permanent flushing with modified atmosphere before the individual packs are sealed.

Gas Mixers and Meterers

In the packaging process the air inside the package is replaced by a gas or a gas mixture. Pre-mixed modified atmospheres are available in different mixtures and under several brand names. Today, in most cases on-site gas mixers are used to create these gas mixtures. MAP gas mixers provide verified gas quality and safety in the packaging process – for germfree and long shelf life food. But above all they offer high flexibility to the user. At the push of a button different mixtures can be produced in the shortest time on one packaging line, depending on the requirements of the product. WITT offers gas mixing and metering systems for all packaging machines used in the food industry, no matter whether it's vacuum packaging, thermos forming, flow pack or chamber packaging machine. The gas mixing systems are adjusted to the specific product type and processes, and require only basic installation requirements.

Gas Analysers

Gas analysers are essential for quality control in the MAP process. The monitoring can be done as continuous analysis during the packaging process or batch sampling after the packaging process. For continuous analysis, a gas analyser module is integrated into the gas mixing system. The gas analyser monitors the correct composition of the gas mixture. Sample testing is part of the quality control system of almost every company working with modified atmospheres. Via a needle, a sample is taken from the package. High quality gas analysers use modern sensors, very precise and rapid and requiring a very low gas volume. They are therefore also suited to packs with very small headspace, and a very low volume of gas inside the package. All data is logged and can be archived for complete QA documentation.

Leak Detection

Modified atmospheres perform only if the protective gas remains inside the package. The package has to be fully leak tight. As a freshness guarantee to retailers and consumers, package leak detection can also provide competitive advantage. Leak testing prevents needless returns, loss of prestige, legal consequences and, worst case, loss of business. To optimise quality assurance, the user can choose between solutions for sample or in-line testing – based on CO_2 or a water bubble test. Package leak detection systems reliably detect even the smallest leak, and are easy to operate. Furthermore, all tests can be digitally logged and documented for customers.

Ambient Air Monitoring

Gas monitoring systems for ambient air protect employees and make the use of gases such as carbon dioxide safer. This is not toxic but accumulates unnoticed in closed rooms and replaces the oxygen in the air. A concentration of 0.3 percent carbon dioxide in ambient air can be a health hazard. The allowed maximum concentration in the workplace is 0.5 percent. At five percent, headaches and dizziness may occur; eight percent and more leads to unconsciousness or even death. The gas level warning unit permanently monitors the concentration of the respective gas in the ambient air, and activates an acoustic and visual alarm when individually definable limits are exceeded. Simple and effective.

For food and vegetables, controlled atmospheres are not just used in packaging but also for the control of ripening control, in special ripening chambers with the help of ethylene. By using gas analysers, the ambient atmosphere can be monitored.

6

Package Testing

Package testing ideally involves the examination of the package contents, levels of packaging, packaging materials, etc. through various qualitative and quantitative procedures. The topics covered in this chapter address the different types of package testing, such as thickness testing, moisture vapor transmission rate, oxygen transmission rate, oxygen scavenger and carbon dioxide transmission rate for a complete understanding.

Package testing (also commonly known as distribution testing, pre-shipment testing and transit testing) is the simulation of real life supply chain physical and climatic events/ hazards, under controlled conditions in the package testing laboratory environment.

Thickness Testing

Background: Many containers for packaged food products and many carbonated beverage bottles contain a gas barrier layer between two layers of structural plastic such as polyvinyl choride, polycarbonate, or polyethylene. The thin barrier layer, which is usually made of ethylene vinyl alcohol (EVOH), polyvinylidence choride (PVDC), polyester, acrylic copolymer, or similar material, is designed to preserve freshness and lengthen shelf life by preventing the migration of gas in or out of the package, for example to keep oxygen out of a frozen food package or to keep carbon dioxide inside a beer bottle.

Equipment: Because the barrier layers in finished containers are very thin, it is generally necessary to make this measurement with a high frequency pulser/receiver system that is capable of working at frequencies of 100 MHz or higher. Minimum measurable thicknesses will always depend on the acoustic properties of specific materials, but barrier layers down to at least 0.001 in. (0.025 mm) in thin-wall containers can typically be measured with this type of system. High frequency delay line transducers in the range 50 MHz to 225 MHz are typically recommended for this type of test.

Typical Procedure: The sample waveform below shows a 100 MHz measurement of a three-layer microwave dinner tray comprising a 0.167 mm (0.006 in.) outer structural layer, a 0.054 mm (0.002 in.) barrier layer, and a 0.271 mm (0. 01 in.) inner structural layer. This test was performed with a 100 MHz V2012 transducer driven by a pulser/receiver.

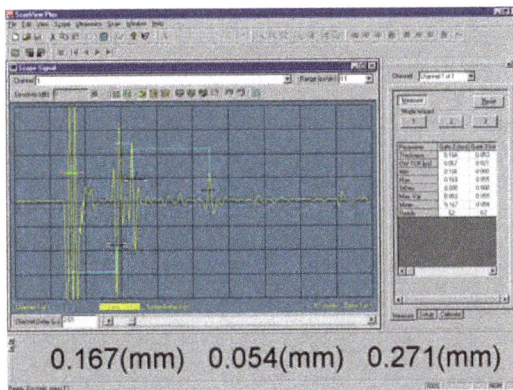

0.167(mm)   0.054(mm)   0.271(mm)

The minimum measurable thickness in a given application will be determined by the highest frequency that is transmitted by the plastic in question. Some plastics are highly attenuating to high frequency sound waves, and thus very thin barriers embedded in thick structural plastic may not be measurable because of low pass filtering effects. For a given product, the measurable thickness range will typically be determined by experimentation with representative samples.

The reflection ratio at the boundary between any two materials is determined by the relative acoustic impedances of those materials. Because virgin and regrind plastics of a given type have essentially identical acoustic impedances, it is not possible to separately measure regrind layers. Also, adhesive layers adjacent to barrier layers are generally too thin and/or too closely impedance matched to measure with ultrasonic techniques and usually cannot be resolved.

As with any ultrasonic thickness measurement, accuracy is dependent on proper sound velocity calibration. Velocity calibration must be performed for each material being measured, on samples of known thickness.

Application: Measurement of individual layers and total thickness in multilayer plastic food and beverage containers.

Moisture Vapor Transmission Rate

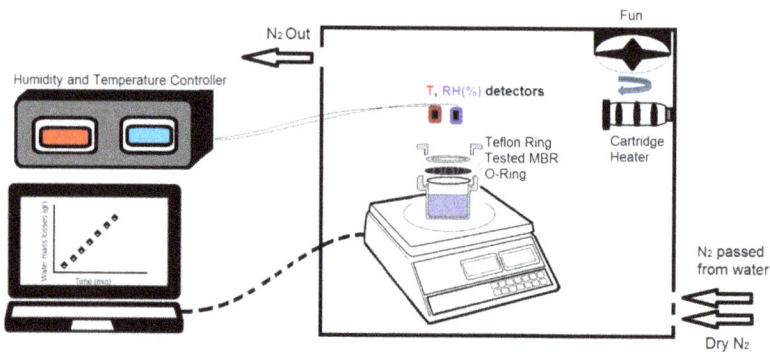

Many packaged products are sensitive to moisture therefore control of water vapor into or out of the package is critical to the contained product's quality. The permeability of packaging materials has a direct effect on a packages performance in terms of the contained products shelf life. This permeation by moisture is measured by WVTR (water vapor transmission rate) or MVTR (moisture vapor transmission rate). The primary purpose of packaging is to get the packaged product to the customer in one piece and suitable for consumption. Packaging plays many roles, containing and protecting products during distribution, storage, and use, while also promoting, educating, and selling.

The barrier property requirements of a particular package are determined by the characteristics of the packaged product and its planned end use. Shelf life is the length of time that packaged perishable, products including foods, pharmaceuticals, and chemicals, etc. are protected and may be still considered suitable for use. Knowledge of the barrier properties of packaging materials is necessary in order to provide a proper package, and an estimate of a contained product's shelf-life. Packaging materials may be permeable to gases, liquids, organic vapors, and water vapor.

(O_2) oxygen and water vapor get the most attention as they can readily move either from the inside of a package to the outside environment, or from the outside environment to the inside of the package. Either way they may have a detrimental effect on the quality and shelf life of a packaged product.

WVTR (water vapor transmission rate), referred to alternatively as MVTR (moisture vapor transmission rate), stands for the standard measure of the passing (permeation) of gaseous H_2O through a substrate. It is the steady state rate at which gaseous water vapor passes through a substrate at specific temperature and relative humidity (RH) conditions over a period of time. Units of measurement are $g/m^2/24$ hr. or $g/100$ $in^2/24$ hr.

Considering certain packaged food products, the control of moisture permeation has a direct effect on product quality and shelf-life. Some packaged wet food products such as meat, seafood, pet food, cheese, muffins, etc., necessitate maintaining a certain moisture level inside a package. On the opposing side, other packaged dry food products including cereals, chips, snacks, pet foods, etc., require a dry, moisture-free level to maintain taste, texture and overall product quality. Products generally will gain or lose moisture rapidly without protective packaging, seeking equilibrium with environmental (the relative humidity or RH). An unsuitable permeable package for example, would likely give the unacceptable and uneatable no longer crisp chips and hard, dried out previously chewy cheese.

Factors affecting WVTR/MVTR values are as follows:

- Thickness of barrier material
- Composition of resin
    ○ Crystallinity/density
    ○ Chain length & distribution
    ○ Chain orientation
    ○ Molecular weight distribution
- Polymer blend
- Coatings
- Additives
- Processing.

Oxygen Transmission Rate

OTR (oxygen transmission rate) is the steady state rate at which oxygen gas permeates through a film at specified conditions of temperature and relative humidity. Value are expressed in $cc/100$ $in^2/24$ hr in US standard units and $cc/m^2/24$ hr in metric (or SI) units. Standard test conditions are 73° F (23°C) and 0% RH.

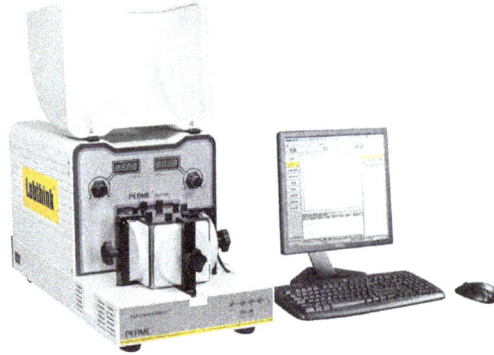

The air we breathe is about 21% oxygen and 79% nitrogen, with very small concentrations of other gases like carbon dioxide and argon. Essential to human and animal life, oxygen gas is also a reactive compound that is a key player in food spoilage. Most of the chemical and biological reactions that create rancid oils, molds, and flavor changes require oxygen in order to occur. So, it is not surprising that food packaging (and some non-food packaging for products where atmospheric oxygen causes harm) has progressed and found ways to reduce oxygen exposure and extend the shelf life of oxygen-sensitive products.

There are two methods for reducing product exposure to oxygen via flexible packaging.

1. MAP (modified atmosphere packaging) is a process for replacing the air in the headspace of a package with another gas before the final seal is made. This is also called gas flushing. The most common replacement gases are nitrogen or nitrogen/carbon dioxide mixtures. The shelf lives of potato chips, dried fruits, nuts, and shredded cheese are commonly extended by this packaging method.

2. Vacuum packaging is where the atmosphere is drawn out and eliminated, rather than being replaced by another gas. This vacuum forces the flexible material to conform to the product shape. Meats (fresh and processed) and cheeses are commonly packaged this way.

OTR is most affected by the following factors:

1. Thickness of barrier layer: Generally, the thicker the oxygen barrier-providing layer, the better the barrier. But there are process and cost limitations that restrict the thicknesses that can be realistically produced or successfully utilized.

2. Copolymer ratio, plasticizer content, and polymerization process: All PVdCs (or EVOHs or PVOHs) are not created equal. Properties are compromised during polymer and product development, so that total performance in target applications is optimized. There can be substantial differences in OTR values depending on the selections made. For example, both ASB-X and AXT are PVdC-coated and sealable, but their OTRs are 4.5 cc/100 in²/24 hr and .40 cc/1 00 in²/24 hr, respectively. ASB-X has the poorer OTR, but a broader seal range than AXT.

3. Base film surface compatibility: The physical and chemical characteristics of the base film surface have a major effect on the OTR after metallization, and to a lesser degree, after coating. This is evidenced by Met PET's exceptional barrier, as well as the difference in OTRs between various metallized OPP products

Oxygen Scavenger

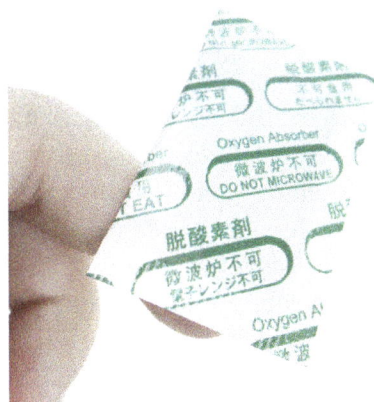

Oxygen scavengers are mainly used for food and pharmaceutical applications, but can also be used for any product that needs a low oxygen storage atmosphere. Essentially, oxygen scavengers are so named because they preferentially absorb oxygen within the environment, thus, preventing the oxygen from reacting with the product. Many other terms have been used to describe oxygen scavengers, which include the following: antioxidants, interceptors, controllers, and absorbers. According to Brody, the definition of an oxygen scavenger is a material in which a chemical (or combination of reactive compounds) is incorporated into a package structure and may combine with oxygen to effectively remove oxygen from the inner package environment. The purpose of an oxygen scavenger is to limit the amount of oxygen available for deteriorative reactions that can lead to reduced functionality of the product. For foods and pharmaceutical products, deteriorative reactions include lipid oxidation, nutritional loss, changes in flavor and aroma, alteration of texture, and microbial spoilage. Typically, oxygen scavengers are used in packages that have air tight seals and are used in conjunction with other means of preservation, such as chemical preservatives, reduced water activity, reduced pH, vacuum packaging, or modified atmosphere packaging.

Applications

Current use

Oxygen scavengers were initially used by the U.S. military for the meals ready to eat (MRE) rations. Foods with oxygen scavengers for MREs include the following: white and whole wheat bread, pound cake, fudge brownies, wheat snack bread, potato sticks, chowmein noodles, nut raisin mix, pretzels, waffles, and hamburger buns. In the retail market, typical uses in the United States include fresh pasta, beef jerky, pepperoni, beer (cap liner and some bottles), ketchup, juice, case-ready meats, prepared foods, and shredded cheese. Use of oxygen sachets is more prevalent and diverse in Japan for products such as seasonings, cheese, aseptically packaged cooked rice, dry pet foods, cough capsules, plant growth hormone, antibiotics, vitamins pills and tablets, medical kits to preserve reagents, kidney dialysis kits, and other products.

Meat Packaging

Oxygen scavengers are used in meat packaging to control color in red meat packages. When the pigment for red meat (myoglobin) is exposed to oxygen, the meat appears as a bright red color, which is associated with freshness and overall acceptability. However, when meat is exposed to too much oxygen over time, the pigment converts to metmyoglobin, which appears as a brown color and is not visually acceptable to consumers. Many studies have used oxygen scavengers in modified atmosphere packages to extend the display life of red meat packages. The conditions under which commercially available scavengers with an oxygen absorbing capacity of 200 mL/sachet could prevent transient discoloration of nitrogen flushed ground beef were studied. The rate of oxygen absorption decreased with decreasing oxygen concentration when the oxygen concentration was between 10% and 20%, but the rate of oxygen absorption became exponentially proportional with time when the oxygen concentration was less than 1%. Oxygen concentration in packages needed to be reduced below 10 ppm in 30 min at $2°C$ or 2 h at $1.5°C$ to prevent transient discoloration. It was determined that to achieve this, more sachets than were economically feasible would have been required at that time. Continued work on prevention of transient discoloration of beef was done by Tewari and others, who determined that steaks removed from a controlled atmosphere masterpack had less discoloration when packaged with an oxygen scavenger and that the number of sachets had more impact on discoloration prevention than the type of scavenger used. The same research group also studied the effect of oxygen scavengers inside retail trays, lidding and over-wrapped trays, which were all equally effective for extending the acceptable shelf life of modified atmosphere, display ready beef, and pork cuts. They also found that eight oxygen scavengers with a capacity sufficient to achieve an oxygen half life of 0.6–0.7 were needed when oxygen concentration could otherwise remain at less than 500 ppm during storage. The redness of meat 96 h after removal from a gas-flushed mother pack was measured. It was found that redness of meat packaged in a retail tray containing oxygen scavengers was better than retail trays that did not contain an oxygen scavenger. In another study, researchers found that a tray combined with a controlled atmosphere mother pack (outer bag), double-flushed with 50% carbon dioxide and 50% nitrogen was effective for maintaining the display shelf life of variety of red meat cuts. The oxygen scavenger helped maintain a low level of oxygen (0.1%) to prevent formation of metmyoglobin.

Another study involved vacuum-controlled atmosphere packaging with carbon dioxide, carbon dioxide flushed packages containing iron-based oxygen scavenging sachets, and packages that contain oxygen scavengers alone for beef stored for up to 20 weeks at −1.5°C . Beef packaged with oxygen scavenger alone provided the best results with regard to drip loss, microbial, and sensory properties. Fresh pork sausages packaged in a 20% carbon dioxide, 80% nitrogen atmosphere with an iron-based oxygen scavenging sachet, were found to have had reduced psychrotropic aerobic counts and extended shelf life with regard to color and lipid stability for 20 days at 2°C. Catfish steaks were packaged in barrier film and vacuum packaged with and without an iron-based sachet. Shelf life was extended 10 days (20 days with sachet vs 10 days without sachet) with the aid of the sachet based on sensory, microbiological, and volatile base nitrogen analysis. The oxygen in the packages containing the sachets reached 0.42% within 24 h of packaging. Labels are manufactured that absorb 10–20 mL of oxygen, and larger labels are starting to become available that scavenge 100–200 mL O_2. Besides labels, oxygen scavengers can also be incorporated into the polymer film. The film is capable of reducing oxygen in the headspace to less than 1 ppm in 4–10 days for products such as dried, smoked meat, as well as processed meat products.

Bakery Products

Baked goods such as bread, pastries, cakes, and cookies can have an extended shelf life with use of oxygen scavengers combined with modified atmosphere packaging. The low-oxygen condition retard molds and other spoilage bacteria and also reduces lipid oxidation, which can produce off flavors. An advantage over using an oxygen scavenger system compared to modified atmosphere packaging alone was because of the fact that lower oxygen levels could be achieved, and oxygen can be reduced in the case of leakage through defective seals. The disadvantages could be cost and possible consumer objection to a packet inserted inside the package that could become loose and cause accidental ingestion by a child or pet. A few studies have reported on the effectiveness of oxygen scavengers for bakery products. Sponge cakes (0.8–0.9 aw) were packed in a modified atmosphere package with oxygen absorber sachets of two different absorption capacities (100 and 210 mL). The cakes were analyzed for mold growth over a period of 28 days at 25°C storage. Modified atmosphere package alone provided some benefits regarding mold prevention; however, combining oxygen scavengers (using either 100 or 210 mL) with modified atmosphere (30% CO_2) prevented mold growth entirely during the 28 days of storage. The results also indicated that a greater benefit existed for cakes with a higher aw (0.9) compared to lower aw (0.8).

In another study, oxygen absorbers were added to wheat crackers formulated with high levels of oil for storage in hermetically sealed cans used as military rations. The study included storage at 15, 25, and 35°C. Shelf life was assessed using sensory panels as well as hexane concentration and headspace oxygen measurements. As storage temperature increased, headspace oxygen decreased within the can. Overall, cans of crackers without oxygen sachets reached unacceptable levels of rancidity within 24 weeks at 25 and 35°C. Cans of crackers with oxygen sachets did not have rancid odors after 44 weeks of storage, regardless of storage temperature. Thus, shelf life of canned crackers was extended for 20 weeks with oxygen absorbers added to the can.

Other Products

Orange juice contained in aseptic packages with an oxygen scavenging barrier layer was found to have better retention of ascorbic acid than packages with plain oxygen barrier film. Packages that contained oxygen scavengers had less mold growth on cheddar cheese compared to packages without the oxygen scavengers over a 16-week refrigerated storage period.

Milk processed using ultra-high temperature processing, was packaged and stored in aseptic pouches that contain an oxygen scavenging film or a pouch without the scavenging film. The milk in the oxygen scavenging pouch had significantly lower levels of dissolved oxygen and levels of volatiles associated with staleness. Hazelnuts were packaged under controlled atmosphere conditions with and without iron-based oxygen sachets. The nuts packaged using the sachets were significantly less oxidized compared to the nuts without the sachets. However, when the sachets were analyzed for volatile compounds, other flavor compounds were also scavenged by the sachets.

Carbon Dioxide Transmission Rate

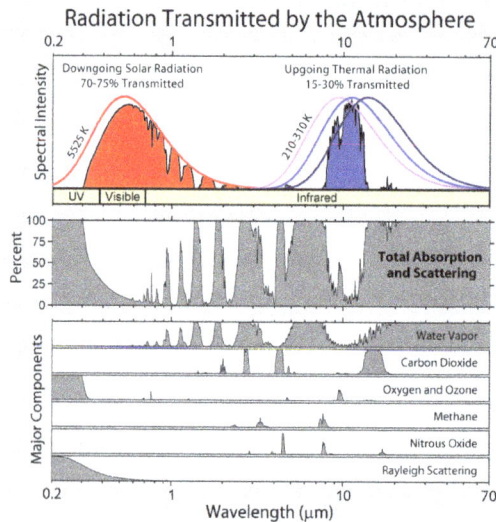

Carbon dioxide transmission rate (COTR) is the measurement of the amount of carbon dioxide gas that passes through a substance over a given period. It is mostly carried out on non-porous materials, where the mode of transport is diffusion, but there are a growing number of applications where the transmission rate also depends on flow through apertures of some description.

References

- Package-testing, distribution-testing: smitherspira.com, Retrieved 28 June 2018

- Thickness-measurement-multi-layer-plastic-food-containers, applications: olympus-ims.com, Retrieved 11 March 2018

- Moisture-and-water-vapor-transmission-rates-in-packaging-458: corkindustries.com, Retrieved 31 March 2018

- Flexographic-otr: polyprint.com, Retrieved 25 May 2018

- Carbon-dioxide-transmission-rate: wikivisually.com, Retrieved 21 March 2018

Package Design

Package design is an integral part of the food packaging process. It involves the identification of the requirements of marketing, shelf-life, structural design, logistics, graphic design, etc. This chapter discusses in extensive detail the different types of package design and the basics of milk package design.

In marketing literature, packaging is a part of the product and the brand. A product's package represents its characteristics and communicates the product information. For consumers, the product and the package are one and the same when they see it on the supermarket shelves. During the purchasing decision, the package assists the consumer by creating the overall product perception which helps the evaluation and the making of the right choice. Furthermore, the package is the product until the actual product is consumed and the package is recycled.

The package design adds value to the package and to the product respectively. Design elements such as colors, font, text, and graphics have an important role in package appearance. Pictures on the package in form of attractive situations can assist in triggering lifestyle aspirations. At the point of purchase, the primary role of the package and packaging design is to catch the consumers' attention and to stand out among the competition in the store or at the supermarket.

Successful package design and packaging itself is the result of the involvement and the work put forth by marketers, designers, and customers. Hence, packaging is a major instrument in modern marketing activities for consumer goods. Prone deems that the package can attract the customers' attention, communicate company's name and image, differentiate the brand from competitors, and enhance the product's functionality. Therefore, the package itself acts as a decisive communication tool and provides consumers with product-related information during the buying decision process.

There is a term that has its origins in packaging and packaging design - product positioning. Positioning recognizes the importance of the product and the image of the company and it is required to differentiate the product in the minds of consumers. In other words, positioning assists the package and product awareness, keeping it present in the consumers mind against competitors in terms of attributes that the brand or company name does not offer. Maggard deems that product positioning induces marketing mix where the elements such as pricing policy, place, products and promotion are included. These elements help to reach the consumers and define the appropriate product positioning in their minds.

Positioning may include different elements which depend on the positioning strategies. This can be global, foreign, and local consumer culture positioning where the attributes such as design, package and performance can have different functions and purposes. However, the main goal of positioning is to provide a successful presentation and explanation on why the consumers should buy a particular product. Therefore, the package and packaging design aims at consumers' attention, whereas the positioning helps the company to place the products properly in the market.

Packaging Functions

Packaging has many functions in different departments. It has its most essential roles in logistics and marketing due to the fact that these two units are strongly connected to the end-users of the product. The task of the package is to sell the product by attracting attention and to allow the product to be contained, utilized, and protected.

Logistical Function

The functionality is a correspondence of packaging to its practical purpose. The roles the package fulfills are related to psychological function, where the package interacts with the consumer and to physical property of a package on a stage of production and product preservation.

1. To contain

    The aim of the package here is to achieve integrity. It means that the product stays in the same condition and does not change its basic form and use, due to the influence of external factors.

    The task of containment is ongoing throughout the product life cycle, from production to the end user and customer. The package function 'to contain' is convenient and beneficial to the consumer as it increases consumer confidence in the contents of the package and the product.

2. To protect

    Protecting the product is a key function of packaging. The protection task is performed not only for physical factors such as transit, but also for environmental influences – moisture, gases, light, temperature, and other.

    Here, the package choice depends on the nature of the goods, distribution and types of hazards it will encounter. Some of the benefits this function can provide for a product are extended shelf life and freshness.

3. To identify

    The role of identification is to provide the consumer with information about the product. Product identification has a description of the contents and consists of product use and legally required information. To some extent, this function can have a promotion role that stimulates the desire to purchase a product and can also assist product branding.

Marketing Tool

Product design is an important marketing variable. It is also a vital instrument in modern marketing activities for consumer goods. To be successful in today's increasingly competitive marketplace, the product design, namely appearance, should include the preferences of consumers. Packaging provides an attractive method to convey messages and information about the product attributes to customers .

Bloch (1995) says that the importance of product design is crucial to the success of a product. It ensures consumer attention for the product, communicates information, and it provides sensory stimulation. According to Berkowitz, an exclusive and unique package design is a way for a new product to be noticeable among familiar packages offered by competitors.

The design of a package contributes to the communication of value and has a strong influence on sales of a particular product. The package and package benefits are essential instruments in marketing strategies.

Packaging as a Decision Making Instrument

Packaging plays a critical role in the purchasing decision. Silayoi and Speece deem that in cases when the consumer is undecided, the package becomes a vital factor in the buying choice because it communicates to the consumer during the decision making time. The way how the consumer perceives the subjective entity of a product through communication elements conveyed by the package, also influences the choice and is the key factor for successful marketing strategies.

Murphy indicates the importance of package design and its influences on consumer decision making process. Murphy distinguishes a two-step decision process the consumer follows during shopping for convenience-packaged products. First step is to decide to examine the product carefully after finding it on the supermarket's shelf. Here, the package design has the power to initiate consumer examination of the product. The second step includes direct experience with the product where the package becomes a "salesman". Hence, the package and packaging design are involved in the consumer selection and purchasing intent.

Analytical and Emotional Decision Making

The functionality of a package is one of the most important areas of packaging design. It has started from simple product identification and has moved to creating branding and communicating imagery in powerful and interesting ways. This communication starts at the point of purchase where the buyers begin to make their choices based on several criteria such as product category, product variety, product size, quantity or volumes, influence of advertising, and many others. All these criteria are dependent on time, browsing or product comparison. However, if none of these factors take place, the purchasing decision will be partly analytical and partly emotional.

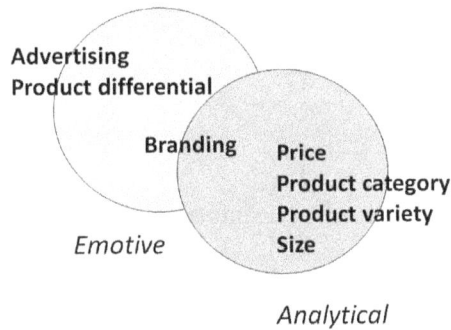

Advertising
Product differential

Branding   Price
Product category
Product variety
Emotive   Size

Analytical

Figure: The purchasing decision is part analytical and part emotional

The analytical part of decision making can be seen when the potential buyers are making a shopping list on which immediate needs are included. The emotional part appears when the buyers need to decide which exact product among its product group they would like to purchase. This choice can be influenced by packaging which can differentiate as "original and best", "just as good but cheaper", or "new and different". Hence, the role of packaging design is to initiate an emotional dialog with the potential purchasers.

Packaging as a Communication Instrument

Packaging design impacts the consumer at the point of sale as well as at the point of future handling and using the product. It becomes a part of the consumers' experience and influences the future purchasing decisions. The way the package can be opened and closed, the way it fits neatly onto the refrigerator, all these factors and qualities can provide emotional feedback which reinforces the brand value and assists product satisfaction.

The packaged product communicates not only through its appearance elements but even more through the overall experience with the whole package. The packaging design includes many features that give the complete picture of the product. Kupiec & Revell suggest that consumers' intention to purchase is dependent on the degree to which consumers suppose that the product will satisfy their expectations about its use. Therefore, the task of package communication is to deliver the right message in order to meet the buyer's needs and emotional desires for purchase.

According to Nancarrow and Wright and Brace, in order to achieve the communication goals and objectives efficiently and to optimize the potential of packaging, companies and manufacturers of fast moving consumer goods need to take into consideration consumer response to the packages they produce, and to integrate the perceptual processes of the consumer into design.

Silayoi and Speece suggest that marketers and designers need to consider consumers past experiences, needs, and wants; understand how packaging design elements get customers attention to the product and get them to notice message on the package; and evaluate packaging design and labeling their effectiveness in the communications effort.

Packaging Elements

The packaging design features and characteristics can highlight and underline the uniqueness and originality of the product. A well designed package sells the product by attracting attention and through positive communication.

Silayoi and Speece based on the review of the relevant literature, define that there are four core packaging elements which affect a consumer's buying decision. These elements are divided into two categories: visual and informational elements. The visual elements include graphics and size/shape of packaging. Informational elements consist of product information and information about the technologies used on the package.

Visual Elements

When creating a package design, it is important to remember that consumers evaluate packaging in different ways. Customers' attitudes towards the package depend also on the process of interconnection between person and package. Here, the level of this involvement influences on the product continuum where the product name varies from high involvement to low involvement product. The difference between them is that the first one has a more substantial effect on the consumer's lifestyle, while the second is less significant and can be habitually purchased. The decision making for high involvement products is less influenced by image issues. Grossman & Wisenblit say that the decision making for low involvement products includes the evaluation of packaging design attributes which is less important, while the graphics and color become more valuable and noticeable. Kupiec et al. suggest that the consumer behavior towards the low involvement products can be influenced by the development of the marketing communications which includes image building.

Graphics and color

Graphics

Graphics include image layout, color combination, typography, and product photography. The combination of all these components communicates an image. Graphics on the package are telling detailed information about the product. It becomes a product branding or identity, followed by the information.

According to Herrington & Capella , when the consumers examine packages in the supermarket, the differential perception and the positioning of the graphics can be the difference between identifying and missing the product. However, eye-catching graphics make the product stand out on the shelf and attract the consumers.

Graphics can affect through colors and printed lines on the package on which different signs and symbols are located. Holograms and combinations of various materials can encourage consumers to touch the package, thereby inspiring them to try the product.

Color

Cheskin says that the selection of the colors and color combinations is a necessary process for creating a good design package. Color is a key element of design due to the fact that it is usually vivid and memorable. The package color can have a significant effect on consumers' ability to recognize the product, the meaning conveyed by the package, its novelty and contrast to other brands and company's names. The package color can be modified without changing the costs, product characteristics and functionality.

Packaging applications have many color-coded messages which are associated with the particular product category. Garber & Hyatt & Starr Koch & Koch, say that in case of food package, color can influence product expectations and perceptions.

Product packages in similar colors may attract attention by means of brand or product category. Dissimilar or novel colors may attract and be preferred by those customers who like novelty. The right choice of colors is an important factor in creating the impression needed to influence brand and product selection.

Size and Shape

Packaging size and shape are also significant factors in designing the package. A consumer interacts with these two elements in order to make volume judgments, e.g. consumers perceive more elongated packages to be larger.

Packaging sizes depend on the different involvement levels. The low involvement food products have a low price which is generated through cost savings created by reduced packaging and promotional expenses. The effect of package size has a strong influence on the purchasing choice when the quality of the product is hard to determine. Therefore, the elongated shape and appropriate size causes the consumer to think of the package as having better product volume and cost efficiency.

Informational Elements

Product Information

Communication of information is one of the core functions of the packaging. This helps customers to make the right decisions in the purchasing process. Coulson gives an example of information significance using a food labeling case: the trend to consume healthy food has emphasized the importance of labeling, which gives the consumer the opportunity to consider alternative products and to make an informed product choice.

Packaging information can create contrary results. It can lead to misleading or inaccurate information through small fronts and dense writing styles which are used on the package. Hausman suggests that experience makes consumers select prospectively the product and it, however, restricts the area of their choice. Hence, the purchase decision making factor depends on the interconnection between information and choices. Here, consumer involvement also takes place. Vakratsas & Amber tell that low involvement includes inattentively reading and examining product information, while high involvement consists of careful evaluation of information and may lead to purchase intentions.

Technology Image

McNeal and Ji deem that the role of packaging in marketing communications is implemented by developments in technology. Here, the technology creates the packages according to trends and consumers' attitudes and behaviors. The role of technology is to meet consumers' needs and requirements. As far as the technology is a communication element, it should be presented visually and, therefore, it will catch more attention and be convenient for consumers.

Packaging Elements Towards Product

The actual package can be considered as a part of the product since it can assist a product's benefits and be important for the product usage. For different products, the shape of the package is the crucial factor for success in the marketplace, whereas size and color can be vital for other goods. Graphics and technology image are the other elements that also contribute to a successful package. Since the packaging is the last marketing communication tool the company uses before the purchase decision is made, the importance of package is highlighted in the communication mix of a company.

The combination of shape creativity and color together with well-designed graphics forms the package and creates consumer emotional appeal. Here, the logistics and marketing aspects are considered and performed in cost efficient way. A company's stability and profitability are dependent on its product relevance and business performance.

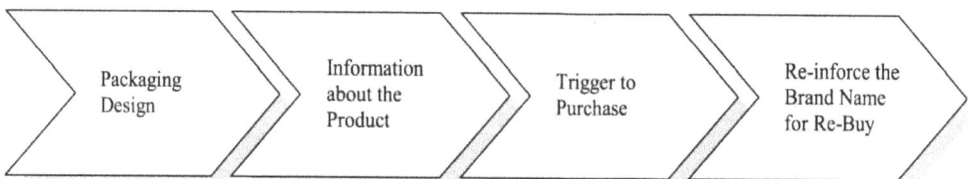

Figure: Packaging design and trigger to purchase

In figure above the process of the package influence on the product consumption is presented. The first thing that evokes consumer attention is product design. By examining the visual elements, a consumer investigates the content and information about the actual product. The package, covered by different design elements, may evoke the desire to purchase the product. When the purchasing decision is made and the product is bought, the product strength and demand is established.

Packaging as a Quality Measurement

The quality of the package as well as the quality of the actual product are the core elements of purchasing decision making. When the consumer forms an opinion towards the new package, the packaging design variables are highly important. The consumer makes a quality evaluation based on the packaging attributes and the overall package. Here, the consumer may perceive the usefulness of the package and judge the favorability of the new product.

Packaging is a quality measurement for the products. According to Grunert & Beck-Larson & Bredahl, when the consumers view the new package on the shelf, they are usually forced to make a quality evaluation of the product through experience with the package.

Quality judgments are influenced by product and package characteristics. When the package communicates high quality, frequently the consumer assumes the product itself as a high quality item. If the package gives the impression of low quality, the consumer perceives the actual product as a low quality item. Underwood, Klein and Burke suggest that consumers instinctively can imagine how the product looks, tastes, feels, smells, and sounds while viewing pictures and images on the package.

Packages should be exciting and safe and have a high quality at the same time. Food product expectations are created by packaging elements such as labeling and product information. Here, the color element also plays an important role. Colors on the package can be perceived and associated with quality attributes, such as flavor and nutrition. Imram believes that a positive effect can be gained by combination of packaging elements: color, clear packages and incident light. In food service, the food products chosen for display are selected for their color and appearance attributes.

The quality combined with product price can influence the purchase intention. Zeithaml says that the price of lower-priced packaged goods receives less attention than high-priced goods. Schoormans and Robben suggest that the attitude towards the package and expected product quality has influence on the consumer's purpose to buy a low-priced packaged product in the supermarket.

Packaging Design as a Tangible Object

The package design contains visual and sensual attributes which communicate to the consumer. Visual elements relate more to the perception and attractiveness, whereas sensual refer to the physical sensation. The way how the consumer interacts with the object, its surface and material can influence the evaluation of product content and quality. Hence, the designing of the package as a physical object is very important as is the creation of attractive visual elements.

From a physical point of view, a package is a container that directly contacts the product, protects, preserves and identifies it. Vidales Giovannetti identifies three types of packaging. First, prime package is in direct contact with the product. Secondary packaging consists of one or more primary packages and has the role of protection, identification and communication tools. Tertiary packaging consists of two previous packages and its function. The task of the third type of package is to distribute, unify and protect products throughout the commercial chain.

Good package design requires knowledge of materials, their properties, manufacturing methods and conversion process. The materials that can be used for producing packages are wood, paper and board, plastics, glass, metals, and textiles. Here, the choice of material depends on the nature of the product, production process, and equipment. The product shelf life, storage and transit requirements also have an impact on the material choice.

The vast variety of products and goods implies a large amount of different packaging methods. Here, as in the case of the materials, the method of packaging is strongly dependent on the actual product. For instance, wrapping is the method of packaging in which an object is enveloped in a sheet of material. The product which can be wrapped does not suit to the products which are in a liquid or unstable condition. Plastic or glass bottles and jars are used as a package for beverages. Steel canning package is a method of preserving perishable food. Hence, the package varies from the product categories as well as from the materials and technical methods it requires.

At the start of every design project, marketers and designers need to have knowledge of the material categories, limitations and possibilities for a particular type of a package, and its conditions and requirements. When the sensual part of the package is well designed and made in a proper way, it is easier to include visual elements with appropriate and selected information and a message which will attract the consumer and evoke an interest in the product.

Influencing Factors

From the historical facts it is known that the package has been utilitarian. It has had specific attributes and special functions which nowadays are changed or replaced with more convenient and functional elements. There are many internal and external factors that have influenced the package and package design throughout their existence. The result of these developments is presented and can be observed by consumers in the supermarket. Nevertheless, packaging still continues to improve and progress.

Consumer Influences

Consumer behavior is influenced by demographic and lifestyle factors. The consequences of demographic factors are an ageing population and an increasing number of people who are moving and living in smaller households. The changes in household sizes also influence the consumer lifestyle. The number of people eating out, as well as the "healthy eating" and sporting activities phenomena changes the society. Accordingly, due to the significant changes, marketers and designers must adapt the package and package design to the consumer's preferences and needs, as well as the visual perception and satisfaction with an actual product.

Environmental Influences

The environment is an important issue for all business areas. Nowadays, governments, official institutions and international companies around the world pay attention to environmental problems and suggest ideas for solving them. Many countries have introduced legislation and regulations for certain material usage or certain trade practice implementation. The European Union has implemented legislation which requires companies to behave in a manner compatible with environmental conservation. The packaging directive describes the minimization of waste and the amount of recycled packaging material. EU says that the directive introduces important restrictions and promotes energy recovery, re-use and recycling of packaging.

International Influences

Internationalization and globalization have a significant effect on the products and consumer behavior. Due to growth of international trade, many products and services are now offered worldwide. Cateora & Graham & Ghauri tell that the international products and brands are marked in a standardized way, whereas other goods need to be adapted to local requirements and preferences. The role of packaging design in the case of internationalization and globalization is to make the product seem different and innovative.

Logistics and Distribution Influences

New logistics solutions are developed due to packaging, which is also a key factor for adaptation for logistics reasons. Packaging and packaging design is a key factor that drives the development of modern distribution systems of dairy products.

Marketing Influences

Consumers bring ideas and also request new products. In order to satisfy customers' demands and requirements, new solutions and ideas need to be found. New techniques and ideas for creating new designs and higher quality of printing can give packages a more luxurious appearance.

Technology Influences

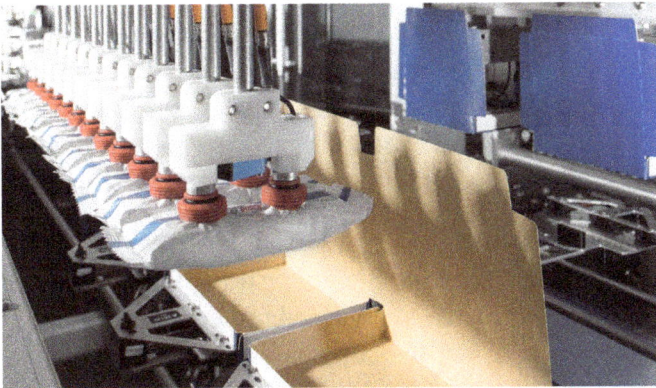

New technology and technological progress in coating and laminating facilitates the enlargement of new materials and combination of materials with better properties. This factor assists the development of new packaging products. Development of printing and printing technology is also a driving factor. Sörensen and Widman say that suppliers of packaging equipments also develop packaging and its design. The development of radio frequency identification technology creates the opportunity for new packaging solutions within distribution systems.

Consumer

Consumer Behavior

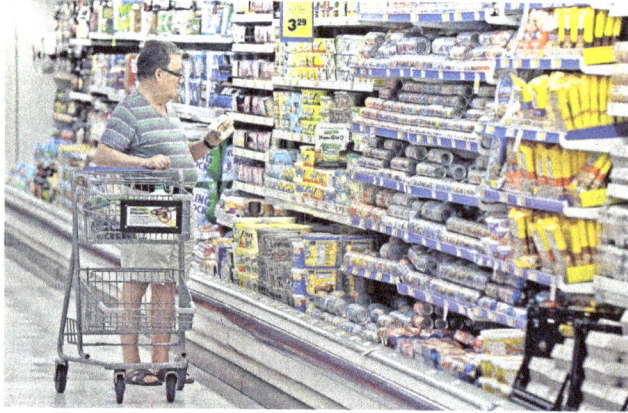

The modern market consists of a big variety and diversity of packages, designs, products, goods, and services. It develops and innovates daily and makes improvements in strategies permanently. However, it would not put so much effort into the development if the consumer and the overall society would not need and require new products, product ideas and functions. The market is the dependable sector of industry and the consumers are only one indispensable element of market performance which allows the industries to exist and grow. In order to create an appropriate product or service, companies need to understand the consumers, their behavior and perception, and to meet their needs and requirements.

Consumer behavior is the process involved when individuals or groups of people select, purchase, use or dispose of products, services, ideas or experiences to satisfy their needs and desires. There are different people with different roles who are involved in this process: the purchaser, whose function is to buy the product or service; the user who uses the actual product or service; and the influencer who provides information and recommendations for or against the product or service without buying or using it.

Understanding the consumer is a good business strategy for the company. The companies and firms operate in order to satisfy the consumers' needs which are the basic concept of marketing. Here, the consumer segmentation is a major element to meet their wishes and requests. The consumer can be segmented by different dimensions such as demographics, geographic, psychographic, and behavioral. Furthermore, there are also different types of consumers who influence the market; e.g. global consumer whose devotion is to brand products and goods and green consumers who feel responsibility for social and moral issues.

Perception

Nature endows people with feelings and senses by which a person can experience the environment. Perception assists a person in understanding his or her surroundings and phenomena as a more detailed concept. In other words, *perception is the process by which physical sensations such as sights, sounds, and smells are selected, organized, and interpreted.*

People during their entire life get tons of information which subsequently is filtered and selected. The information can be in form of natural or background noise, advertising or news, or even a sound. Here, people get information automatically and react on it according to their needs, wishes or experience.

People notice only a small amount of stimuli and pay attention to an even smaller amount. The meaning of these stimuli is interpreted by the individual according to his needs and experiences. Figure represents the process of perception where the stages of sensation, attention and interpretation are illustrated.

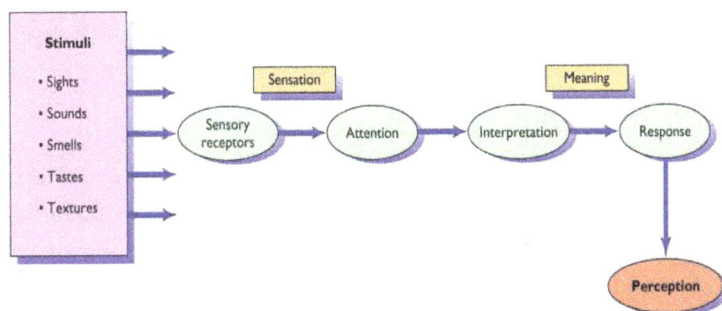

Figure: An overview of the perceptional process

Stimuli

Sights/Colors

Colors are rich in symbolic value and cultural meanings. For instance, the color red is associated with blood, wine-making, activity and heat in many countries but is poorly received in some African countries. The color white is identified with purity and cleanliness in the west while in parts of Asia this color symbolizes death. Yellow indicates a merchant in India. Grey means inexpensive in Japan and China but high quality and expensive in the U.S. Hence, colors can provide different meanings but can also be used to evoke positive or negative feelings. Marketers consider color as an integral part of their strategies. Before introducing color and color combinations, marketers need an understanding of how it is perceived in each part of the world and consider the fact that the popularity of colors is depending on the culture.

The marketers can create the colors which will meet consumers' expectations. The color combination can be associated with a particular brand or company name.

When creating new design packaging or advertising, marketers rely on visual elements and use them in a proper way in order to attract consumers.

Sound

Solomon & Bamossy & Askegaard deem that sound can affect people's feelings and emotions. It is well-known fact that music can affect mood and the influence of speaking rate on attitude change and message comprehension.

Smell

The pleasant smell of a product can affect the emotions or have a calming effect. The smell can evoke memories or relieve stress. The scent can also communicate information about the product: novelty e.g. book or fashion e.g. perfume.

Taste

By this human perception element, people can feel and evaluate product quality e.g. cheese or chocolate. Taste has effects on consumer experience with a product.

Texture/Touch

This sensory channel is very important. A person's mood is stimulated or relaxed by sensation of the skin, for instance. Touching assists consumers in evaluating a material whether it is smooth, soft or stiff and assessing the product quality or fabric texture.

Sensation

The introduction to stimulus is the first step in the processing the information. The sensory organs are activated and are ready to process and evaluate the information. Here, consumers filter and select necessary information and adapt it to their needs and desires.

Attention

Attention is the degree to which consumers focus on stimuli within their range of exposure. Nowadays consumers are exposed to a lot of advertising and design stimuli forcing marketers to become creative and original in products and images in order to appeal to the consumers. When many stimuli are competing to be noticed, one will receive the attention to the extent that it is different from others. Here, size and color can catch the attention and can be a way to achieve contrast.

Interpretation

The final stage in the process of perception is interpretation. Interpretation is the process through which individuals and groups give a meaning to exposed stimulus. People differ in terms of the stimuli they perceive as in meanings and interpretations they can make. Here, the interpretation can also vary from cultural and individual differences.

Perception and Packaging

Food products use many packaging attributes, combining colors, designs, pictures and images, shapes, symbols and signs, messages and information. All these elements can attract and sustain attention. How consumers perceive the subjective entity of products, as presented through packaging and design attributes, influences the choice of purchasing and is the key to success for many packaged food product's marketing strategies.

Types of Package Design

Attractiveness

Product and package communicate via different attributes. The goal of these elements is to catch attention and attract the consumer. Appropriate communication variables are used for different types and groups of consumers in order to meet their needs and desires.

Source of attractiveness is the dimensions of a communicator which increase his or her persuasiveness where expertise and attractiveness are included. Source of attractiveness also refers to the source's perceived social value. Here, the role of celebrities, current trends, and healthy lifestyle create the attractiveness of the product which is presented through the package and design.

Image is one of the elements that can attract and interest the consumer. The general image and good design is central to the perception of the goods. There are different kinds of pictures and images; the decision of the particular type of print depends on the nature of the product and the target group which will investigate and use the actual product. Thus, a provocative picture can be too effective; it can attract so much attention that the consumer is not able to notice or recognize other aspects of the product. The humorous images affect products. This type of communication tool increases the message interpretations and acceptance. Solomon et al. suggest that fear and humor are often used in communicative strategies in the form of storytelling. This occurs due to the fact that the product benefits which are presented on the package or in advertisement are intangible and must be given a tangible meaning by expressing them in a form that is concrete and visible.

The meaning of the picture is strongly related to the product. The message conveyed on the package should correspond to the information the company wants to deliver to potential consumers. It happens that the picture or design is unclear or interpreted in a wrong way by the consumer which leads to a decrease of the product purchasing power and the company's image respectively. Here, the design aspects, as well as cultural differences should be considered and applied in a proper way.

Attention is the degree to which consumers focus on stimuli within their range of exposure. Nowadays, consumers are exposed to many advertising stimuli, marketers become more creative in making an attractive and unique design to increase interest in the products. The upside down printing parts, putting ads in unconventional place, vivid colors, and unusual picture, all these solutions of creating packages and products catch consumers' attention. After a package or product has acquired the attention of a potential customer, the following steps are interest or motivation, involvement, decision making and purchasing desire and ability.

Attractiveness and Packaging

Packaging plays a major role when products are purchased. It is the first thing that the consumer sees before making the final decision to buy. The importance of package design increases with the arrival and popularization of self-service systems. Here, Cervera Fantoni says that packaging is on the foreground in attracting attention and causing the purchase. *Self- service has transferred the role of informing from the sales assistant to advertising and packaging.* Vidales Giovannetti deems that this has become the reason, why packaging is called the "silent salesman" which provides necessary information about the product, its quality and benefits.

The rhythm of life accelerates and the amount of time spent on making choices decreases. People live in a rush, in big cities and are under higher levels of perceived time pressure. Thus, they purchase fewer products than they intend to. Hausman says that *products purchased often appear to be chosen without prior planning and represent an impulsive buying decision*. Here, the package design that attracts consumers at the point of sales assists them in making decisions quickly in the store. The eye-catching package has more opportunities to be noticed and chosen against the competitors and be purchased.

Underwood et al. deem that pictures are more effective than the text when the package wants to stand out and differentiate itself from the competitors' products. Consumers process visual information quickly compared to words.

The package with a strong ethical identity with respect to the environment and human relations, with a unique appearance and a sufficiently different image assists the consumers' decision-making and drives purchasing.

Consumer Decision-making

Consumers make the decisions to buy the products or goods according to their needs and requirements. Since consumers have experience and product knowledge, they tend to make a purchase choice. The decisions are built around several factors and attributes which communicate to consumer through package and product.

The consumer goes through several steps in order to make a purchase. First step is called *problem recognition*. Here, the consumer sees the difference between the current state and the desired one. Second step is *information search*. The consumer investigates the data and makes a reasonable decision. Third step is *evaluation of alternatives*. Here, the consumer collects the alternatives, identifies, categorizes, and compares them against his criteria. Fourth step is called *product choice*. Here, there are two rules that drive the decision. The non-compensatory rule reduces the number of alternatives that do not fit the criteria the consumer has set up. Compensatory rules mean that the consumer considers all alternatives carefully in order to make the right choice.

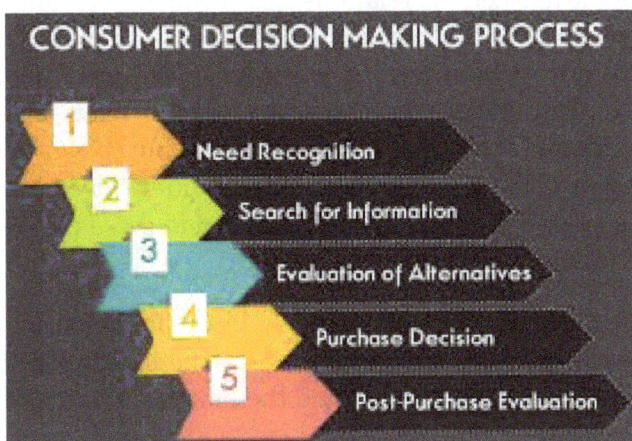

CONSUMER DECISION MAKING PROCESS

1 Need Recognition
2 Search for Information
3 Evaluation of Alternatives
4 Purchase Decision
5 Post-Purchase Evaluation

There are three types of consumer decisions: extended decision-making is initiated by a motive that is fairly central to the self-concept, and the eventual decision is perceived to carry a fair degree of risk; limited decision-making is straightforward and simple action where the motivation is not really to search for information and evaluate alternatives; habitual decision-making is a routine and subconscious activity.

In the end of every decision making process, the outcome shows whether the products satisfy the consumer's needs and wants or not. Hence, the results of the process of investigation and evaluation are presented and if needed can be reviewed or modified.

Consumer Decision-making and Packaging

There are many communication instruments in marketing such as advertising and product demonstration. However, when these traditional tools face the problem of reaching the target audience, the package and packaging design are better able to reach and influence potential and prospective customer. Here, the necessary attributes of the packaging design can become very effective in marketing communications.

The package interaction with the consumer can evoke attention and involve the consumer with the information processing process. Here, the information about the product is investigated and results with the buying decision. However, if the consumer is not motivated to learn about the product, the package characteristics such as color, graphics, image, and shape can induce a positive or negative attitude towards the product.

Therefore, the package standing on the shelf affects the consumer decision making process. The package design needs to insure that consumer response is favorable and that they perceive the actual packaging design positively. Kupiec et al. say that the intention to purchase also depends on the degree to which consumers expect the product and package to satisfy their needs and desires.

Cooperation between Consumer and Company

Since the creation of the market, the relationships between seller and buyer are the core and necessary components of the market performance. The roles of these two market players have changed throughout the years and have made an impact on the modern economy and market. In practice, there are two types of relationships: the marketer sets the trends of goods and services for consumers; the buyer's wishes and requests for new goods and services make markets work on consumers' needs and desires. Hence, the consumer – seller relationship is in equilibrium and these two parties co-exist in a dynamic balance of influence and interdependence.

The consumer behavior is an element that should be predicted by the companies in order to sustain a working relationship with their partners in production and consumption. Consumers, in turn, need to understand the market opportunities in technology, environment, and business.

In case of packaging, cooperation between consumers and business is essential and highly important. According to the consumers' perception, preferences and evaluation, the package design is created and the actual package is launched.

A company needs to make various tests and experiments with the package, packaging design and the customer to ensure that the consumer perceives it positively, investigates it carefully, understands the package usage clearly, and has a purchase intention. The partnership with consumers also helps to develop attractive point-of-sale packaging design which is present on the shelf-ready package. Therefore, involving consumers in the design process is the best way to determine which aspects of packaging need to be improved or developed. Gofman et al. believe that affording consumers a role in packaging creation increases the probability that they will choose the product later on.

The relationship between consumer choices and design characteristics of packaging is a component that marketers of packaged products need to understand in order to develop effective marketing strategies.

Consumer Benefits

Consumers and potential, prospective buyers are the revenue and profit creators. Companies innovate and develop products in order to satisfy and meet consumers' needs. Here, consumer benefits are crucial because they enhance consumer experience in using a particular package and product. The most important benefits are: *ease of opening, closing and releasing; ease of handling; convenient methods of product processing; security and product integrity.*

The main function of the packaging is to consider the end-user and to make the task of opening and using the product as easy as possible. The example of creating consumer benefits in opening can be seen through gable-topped cartons package for milk. Here, the manufacturers create plastic caps and spouts that make the process of opening easy and simple.

The features that provide benefits should be clear to the consumer and should assist him in making the right purchase choice. All types of packaging applications need to communicate directly to the consumer and help him or her to make a decision based on particular package attributes and components.

Milk Package Design

2D and 3D of Package Design

Package design is a merging of two- and three-dimensional design, promotional design, information design, and engineering. The casing aspect of package design is three-dimensional—it's the structural design. It is a form (think carton, bottle, can, jar, tin, wrapper, bag, etc.) made out of materials and substrates (glass, metal, plastic, paper, etc.) possibly involving special finishes. Your project may require designing a new structural form or be for an existing form (think carbonated soda can). On the shelf, packaging is seen from a frontal or 2D point of view. Once taken off the shelf, all sides count; the form is a three-dimensional solution, and each plane of the form relates to every other. When designing the graphics (all the type and images), all surfaces and sides of a pack- age must be considered. Environmental context must be taken into account—for example, light falling on the package's form in a setting. As with any design, it is best practice to solve the graphics and structural form design aspects at the same time, which would make for the most organic solution.

For packaging, if the form is new, then a prototype(s) needs to be constructed. If the form exists, you could start by sketching the face panels, with 2D sketches, and position them in a photo-editing program on a form.

Audio Package Design

For many people, listening to a CD at home involves contemplating the cover, reading the inside booklet and lyrics, and looking at the photographs of a favorite recording artist such as Willie Nelson. We may glance at a superbly designed shampoo bottle with appreciation, but we study a CD cover intently. Looking at a CD cover becomes part of the listening experience. Audio package design can draw in a new listener as well as engage a fan.

People feel very strongly about the music they enjoy and the recording artists they prefer. Audio package design absolutely must reflect the recording artist's or group's sensibility—no equivocations. The package design must express the unique quality of the artist or group, while inviting the person browsing to consider and purchase it.

To understand Package Design better, let's take example : Milk Package

Milk package design is a core element of the current report. Milk is a dairy product which is always needed for consumption and cooking. It is a demanded product all over the world and it is an integral part of food. Hence, milk was chosen due to the reason of purchasing demand. In the market where the high competition exists, it is important to differentiate. In case of milk package, it is crucial to have an attractive design in order to gain consumers' attention and to stimulate purchase decisions.

There are a few companies in Finland which are specialized in the package design creation and the milk package production. In case of the current report, two companies are considered: Valio Ltd and Tetra Pak Ltd. Valio is a company that produces dairy products such as yoghurt, milk, cheese, butter and etc. Valio is the biggest milk manufacturer in Finland that has 86% of milk market shares. The task of Valio is not only to produce dairy products but also to present them in attractive form. The company creates the design, makes images and slogans. However, it does not produce the actual package. This task concerns another company which is also involved into package design creation. Tetra Pak is a food packaging and processing company. Tetra Pak offers packaging and processing solutions for dairy, beverages, ice-cream and prepared food. The company creates and tests different types of package designs, offers new ideas about packages production. Using new technologies and equipment, Tetra Pak produces high-quality packages for different products. Therefore, the milk package design is a result of the involvement and the work put forth by these two companies, Valio and Tetra Pak.

Valio creates milk package designs by working with different designers. For instance, in year 2011 Valio has taken part in the event where several leading Finnish designers have created their own visions of milk for Valio's basic and organic milk cartons. These design images have appeared on the packages in November 2011. On its website, Valio has provided information about the design ideas that have been created by Finnish designers. Illustrator Klaus Haapaniemi's milk carton is decorated with fantasy-world folklore drawings. Designer and interior architect Kristiina Lassus has created meditative patterns in cool colours to represent a picture of calm amidst the daily grind. Architect Jenni Reuter adopts a spatial view of the milk carton and takes us inside an atmospheric barn. (valio.fi). Valio puts a lot of effort to be unique in package design. It invites creative and innovative persons who help to make the package which stands out and attracts consumers.

Tetra Pak produces various packages for different products. Here, milk packages exist in different designs, shapes, and volumes. In Finland, for instance, Tetra Pak produces chilled packages, namely Tetra Rex. The Tetra Rex package gable top carton presents in four formats: Base, Base Plus, Max, Mid and Slim. The gable top packages are available with or without opening devices. The volumes of these packages range from 237 ml. portion to 2000 ml. family packs. The Tetra Rex packages are made from the material that suits the product. The packages have a wider closure and perfect pouring angel what makes it is easy to control the flow of a liquid dairy product. For producing this type of package Tetra Pak uses safe and efficient sterilization technologies. The Tetra Rex packages keep products as fresh and tasty as the day they have been produced.

The package design produced by these two companies has a modern and attractive appearance, it catches consumers' attention, stimulates purchasing desire and motivates during decision making process. The stable and easy-to-use package is made from material that involves consumers to experience and investigate the package and its design. The package guarantees the freshness and taste of the milk. The combination of all these benefits makes the design and package unique and also allows it to stand out among the competition.

Milk package design is created in the same way the company does all other product package designs. Here, the process of package designing includes several steps. First, the knowledge and information about product concepts and customer groups are collected. Then, the company starts planning the concept images and making several versions of the design for consumer testing. The final design version is decided by consumers based on company's preliminary work.

The target groups can have different milk packaging designs. For instance, Plusmilk, Hilja-milk or organic milk can be aimed at different consumer groups. The package design is the most important media of a brand and the design should be created according to the brand personality in each case. Milks are meant usually for the whole family. Some special milk such as Profeel-drinks have narrower target group.

Milk packages have different colors and there is a reason for it. Skimmed milk has the color light blue and semi skimmed milk is presented in the color dark blue. The whole milk packages are usually in color red in Finland. Milk package colors depend on the fat content (%) of milk, however, in branded milk packages the color varies more.

Some package designs stay the same over long time; some of them are changed within a short period of time. However, being the most important media of a brand, package design should keep its key elements in order to be noticed by consumers. Salusjärvi said that "One way to keep the brand alive is to renew the package every now and then. For new products it is, of course, natural to have redesigned package and design whenever it is possible".

Most of the packages for milk are made from carton blanks. However, on the supermarket's shelves consumers can find packages which imitate plastic surface, but originally they are done from carton. Tetra Pak does not use glass or plastic as a material for producing packages.

The milk packages have the shape of cuboids. According to technical issues it is possible to produce only these blanks with cuboids form. From the logistics point of view, it is more effective to produce this type of shapes due to the reason that more packages can be transported using one pallet. All these factors have become a reason why Tetra Pak does not produce packages with a shape of a bottle even if it can be easier to handle.

The milk package design is very important for decision making process. During the purchasing, the package helps the consumer to evaluate the product. The design of the package communicates the information about the product quality. Its material and ease of use can assist consumer in choosing the particular milk product. The size and shape of the milk package are essential elements of packaging design which drive the consumer attention and influence the purchasing decision. The overall product perception is created by the package design having the high value for the consumer.

The packaging design elements have different impacts on the consumer. Some of them catch the consumer attention; some of them stay ignored. However, the findings show that different designs attract with different elements. The milk package designs used in the questionnaire provide the evidence that design elements such as graphics and image, color, product information, and shape of the package play different roles.

The consumer perception and attraction varies a lot. People perceive and evaluate package and its design in different ways: where one likes the image and color, another would not even pay attention to it. The results show that two milk package designs are perceived differently and attraction elements are evaluated according to the particular milk package design. The product recognition is also an element of the perception. Two milk packages are recognized as a milk product by different elements: one by image and another by the name.

Food Labels

Food labels include critical information pertaining to the contents of the food package, its ingredients or information regarding certain allergy risks such as presence of soy or gluten. This chapter gives an analytical view on food labels, labeling policies and government regulations for a comprehensive understanding of the industrial standards of food packaging and labeling.

A panel found on a package of food which contains a variety of information about the nutritional value of the food item. There are many pieces of information which are standard on most food labels, including serving size, number of calories, grams of fat, included nutrients, and a list of ingredients. This information helps people who are trying to restrict their intake of fat, sodium, sugar, or other ingredients, or those individuals who are trying to get enough of the healthy nutrients such as calcium or Vitamin C. The label provides each item with its approximate percent daily value, generally based on a 2,000 calorie diet.

Get your Nutrition Facts Straight

Nutrition Facts
Serving Size 2 crackers (14 g)
Servings Per Container About 21

Amount Per Serving

Calories 60   Calories from Fat 15

	% Daily Value*
Total Fat 1.5g	2%
Saturated Fat 0g	0%
Trans Fat 0g	
Cholesterol 0mg	0%
Sodium 70mg	3%
Total Carbohydrate 10g	3%
Dietary Fiber Less than 1g	3%
Sugars 0g	
Protein 2g	

Vitamin A 0%	•	Vitamin C 0%
Calcium 0%	•	Iron 2%

* Percent Daily Values are based on a 2,000 calorie diet. Your daily values may be higher or lower depending on your calorie needs:

		Calories:	2,000	2,500
Total Fat	Less than		65g	80g
Sat Fat	Less than		20g	25g
Cholesterol	Less than		300mg	300mg
Sodium	Less than		2400mg	2400mg
Total Carbohydrate			300g	375g
Dietary Fiber			25g	30g

The Nutrition Facts food label gives you information about which nutrients are in the food. Your body needs the right combination of nutrients, such as vitamins, to work properly and grow.

The Nutrition Facts food label is printed somewhere on the outside of packaged food, and you usually don't have to look hard to find it. Fresh food that doesn't come prepackaged sometimes has nutrition facts, too.

Most nutrients are measured in grams, also written as g. Some nutrients are measured in milligrams, or mg. Milligrams are very tiny — there are 1,000 milligrams in 1 gram.

Other information on the label is given in percentages. Food contains fat, protein, carbohydrates, and fiber. Food also contains vitamins, such as A and C, and minerals, such as calcium and iron. Nutrition specialists know how much of each one kids and adults should get every day to have a healthy diet. The percent daily value on a food label tells you how this food can help someone meet these daily goals.

On food labels, they base the percentages on a 2,000-calorie adult diet. So looking at the label above for two crackers, a grownup would see that they provide less than 1 gram of fiber, only 3% of the person's daily needs. So that means he or she would have to eat other foods to get 100% of the fiber needed each day. Similarly, the person would see that the crackers provide nothing toward the daily goals for vitamin A, vitamin C, calcium, or iron.

Comparing Labels

Food labels aren't ideal for kids because they're calculated based on what adults need to eat. A kid's diet might be more or less than 2,000 calories, based on your age, whether you are a boy or girl, and how active your are. Likewise, kids may need more or less of certain food components and nutrients, such as calcium and iron.

But kids can still get important information from food labels. They can get a general idea about what the food contains, how much is in a serving, and how many calories are in a serving.

Kids also can use labels to compare two foods. Which one has more fiber? Which one has more fat? Which one has more calories per serving?

The ingredient list is another important part of the label. Ingredients are listed in order so you get an idea of how much of each ingredient is in the food. When something is listed first, second, or third, you know that this food probably contains a lot of it. The food will contain smaller amounts of the ingredients mentioned at the end of the list.

With that in mind, check ingredient lists to see where sugar appears. Limit foods that mention sugar in the first few ingredients. That means it's a very sugary food. Sugar has different names, so it might also be called high fructose corn syrup, corn syrup, sucrose, or glucose.

Serving Size

The nutrition label always lists a serving size, which is an amount of food, such as 1 cup of cereal, two cookies, or five pretzels. The nutrition label tells you how many nutrients are in that amount of food.

Serving sizes also help people understand how much they're eating. If you ate 10 pretzels, that would be two servings.

Servings per Container or Package

The label also tells you how many servings are contained in that package of food. If there are 15 servings in a box of cookies and each serving is two cookies, you have enough for all 30 kids in your class to have one cookie each. Math comes in handy with food labels!

Calories and Calories from Fat

The number of calories in a single serving of the food is listed on the left of the label. This number tells you the amount of energy in the food. The calories in a food can come from fat, protein, or carbohydrate. People pay attention to calories because if you eat more calories than your body uses, you will gain weight.

Another important part of the label is the number of calories that come from fat. People check this because it's good to limit fat intake to about 30% or less of the calories they eat.

Total Fat

The total fat is the number of fat grams contained in one serving of the food. Fat is an important nutrient that your body uses for growth and development, but you don't want to eat too much. The different kinds of fat, such as saturated, unsaturated, and trans fat, will be listed separately on the label.

Cholesterol and Sodium

These numbers tell you how much cholesterol and sodium (salt) are in a single serving of the food. They are included on the label because some people should limit the amount of cholesterol and salt in their diets. Cholesterol and sodium are usually measured in milligrams.

Total Carbohydrate

This number tells you how many carbohydrate grams are in one serving of food. Carbohydrates are your body's primary source of energy. Under this heading, the number of grams of sugar and grams of dietary fiber in each serving are listed.

Protein

This number tells you how much protein you get from a single serving of the food. Your body needs protein to build and repair essential parts of the body, such as muscles, blood, and organs. Protein is often measured in grams.

Vitamin a and Vitamin C

These list the amounts of vitamin A and vitamin C, two important vitamins, in a serving of the food. Each amount is given as a percent daily value. Other vitamins may be listed on some labels.

Calcium and Iron

These list the percentages of calcium and iron, two important minerals, that are in a serving of the food. Again, each amount is given as a percent daily value and other minerals may be listed on the label.

Calories per Gram

These numbers show how many calories are in one gram of fat, carbohydrate, and protein. This information is the same for every food and is printed on the food label for reference.

Use-by and Best Before on Food Labels

Foods with a shelf life of less than two years must have a 'best before' or 'use-by' date. These terms mean different things. The 'best before' date refers to the quality of the food – food stored in the recommended way will remain of good quality until that date.

It may still be safe to eat certain foods after the 'best before' date, but they may have lost quality and some nutritional value. By contrast, foods that should not be consumed after a certain date for health and safety reasons must have a 'use-by' date and cannot be sold after that date. You will find 'use-by' dates on perishables such as meat, fish and dairy products.

Some foods carry the date they were manufactured or packed, rather than a 'use-by' date, so you can tell how fresh the food is. For example, bread and meat can be labelled with a 'baked on' or 'packed on' date.

To be sure that food is okay to eat:

- Check the 'use-by' or 'best before' date when you buy food.

- Keep an eye on the 'use-by' or 'best before' dates on the food in your cupboards. Don't eat any food that is past its 'use-by' date, even if it looks and smells okay.

Lists of Ingredients on Food Labels

All ingredients must be listed in descending order by weight, including added water. Remember that:

- The ingredient listed first is present in the largest amount.

- The ingredient listed last is present in the least amount.

If an ingredient (other than an allergen or additive) makes up less than five per cent of the food, it does not have to be listed. Where there are very small amounts of multi-component ingredients (less than five per cent), it is permitted to list 'composite' ingredients only: for example, it may say 'chocolate' (rather than cocoa, cocoa butter and sugar) in a choc chip ice cream. This does not apply to any additive or allergen – these must be listed no matter how small the amount.

Food Additives on Food Labels

All food additives must have a specific use and they must be assessed. Food additives can be used to improve quality of a food or improve the flavour or appearance of a food. They must be used in the lowest possible quantity that will achieve their purpose.

Food additives are given in the ingredient list according to their class, which is followed by a chemical name or number. For example:

- Color (tartrazine).

- Color (102).

- Preservative (200).

- Emulsifier (lecithin).

The same food additive numbering system is used throughout the world. Vitamins and minerals are also listed under food additives.

The Nutrition Information Panel on Food Labels

The nutrition information panel (NIP) tells you the quantity of various nutrients a food contains per serve, as well as per 100 g or 100 ml. It's best to use the 'per 100 g or 100 ml' value to compare similar products, because the size of one 'serving' may differ between manufacturers.

Under labelling laws introduced in Australia in 2003, virtually all manufactured foods must carry an NIP. There are exceptions to the labelling requirements, such as:

- Very small packages and foods like herbs, spices, salt, tea and coffee.

- Single ingredient foods, such as fresh fruit and vegetables, water and vinegar.

- Food sold at fundraising events.

- Food sold unpackaged (if a nutrition claim is not made).

- Food made and packaged at the point of sale.

Nutrition Claims on Food Labels

Don't be misled by labelling tricks and traps. The terms used are often misleading. For example:

- The term 'light' or 'lite' doesn't necessarily mean that the product is low in fat or energy. The term 'light' may refer to the texture, color or taste of the product. The characteristic that makes the food 'light' must be stated on the label.

- The claims 'no cholesterol', 'low cholesterol' or 'cholesterol free' on foods derived from plants, like margarine and oil, are meaningless because all plant foods contain virtually no cholesterol. However, some can be high in fat and can contribute to weight gain if used too generously.

- If an item claims to be 93 per cent fat free, it actually contains 7 per cent fat, but it looks so much better the other way.

- 'Baked not fried' sounds healthier, but it may still have just as much fat – check the nutrition information panel to be sure.

- 'Fresh' actually means the product hasn't been preserved by freezing, canning, high-temperature or chemical treatment. However, it may have been refrigerated and spent time in processing and transport.

Nutrition Claims and Health Claims on Food Labels Must Meet the Guidelines

Manufacturers can make various claims regarding the content of their product, controls the use of content claims on food labels.

Nutrition content claims make statements about certain nutrients or substances in a food, for example 'high in calcium.'

For a manufacturer to make various claims, their products must meet various guidelines including:

- No added sugar – products must not contain added sugar, but may contain natural sugars.

- Reduced fat or salt – should be at least a 25 per cent reduction from the original product.

- Low fat – must contain less than 3 per cent fat for solid foods (1.5 per cent for liquid foods).

- Fat free – must be less than 0.15 per cent fat.

- Percentage of fat – remember 80 per cent fat free is the same as 20 per cent fat, which is a large amount.

- Good source of – must contain no less than 25 per cent of the RDI for that vitamin or mineral.

Health claims can also be made about a food product and relate to a nutrient or substance in a food, and its effect on health. There are two types of health claims:

General level health claims demonstrate the effect on a health function due to a nutrient or substance that is present in a food. 'Calcium is good for bones' is an example.

High level health claims refer to a serious disease or biomarker and its relationship to a nutrient or substance according to scientific research. For example, diets high in calcium can reduce the risk of osteoporosis.

Food and Nutrition Labeling Policy and Regulations

Food and nutrition labeling requirements impact food formulation and marketing of foods. This web-based content is intended to help food scientists, regulatory, marketing, and other stakeholders stay up-to-date on labeling requirements. It provides links to the labeling regulations of the United States and several other countries, and standards and resources from non-government organizations.

Food labels are designed to provide consumers with information to help make informed food choices. In the United States, food laws such as the Federal Food, Drug and Cosmetic Act (FD&C Act), Food Quality Protection Act, Food Allergens labeling, and Consumer Protection Act impose different labeling requirements on foods and beverages. These laws were passed to prevent consumer deception, ensure fair trade practices, ensure food safety, improve public health, inform about possible health risks (allergen labeling) etc. Under provisions of the U.S. law, importers of food products intending to market in the U.S. are responsible for ensuring that the products are safe, sanitary, and labeled according to U.S. requirements. Similarly, food products exported from the U.S. should meet the packaging, labeling, and other special conditions required by the country.

In the U.S., food label is primarily regulated by the Food and Drug Administration (FDA) and United States Department of Agriculture (USDA), depending on the type of foods and beverages. By law, several elements (listed below) are mandated on the food label, which impact the development of foods and food reformulation/formulation.

- Statement of Identity
- Net Contents
- Nutrition Facts Panel
- Ingredient Statement
- Statement/Warning on Allergen
- Country of Origin
- Name and Place of Business

Food Label Regulations in the U.S.

Recent Developments

On May 27, 2016, the FDA announced the final rules on Revision of the Nutrition and Supplement Facts Labels and Serving Sizes of Foods That Can Reasonably Be Consumed At One Eating Occassion; Dual-Column Labeling; Updating, Modifying, and

Establishing Certain RACCs Serving Size for Breath Mints; and Technical Amendments. The changes in the new Nutrition Facts label include:

- Sugars added to foods will be required on the label in grams and as percent Daily Value (%DV). Added sugars will be declared as "Includes 'X' g Added Sugars" directly under total sugars.

- Sugars will now be declared as total sugars.

- The %DV and the actual quantities of vitamin D and potassium will be required.

- The actual quantities of calcium and iron will be required, in addition to the %DV.

- The Daily Reference Values for sodium will be reduced to 2,300 mg/d from 2,400 mg/d.

- Vitamin A and C are not required on the label, but can be included on a voluntary basis. However, if a specific claim is made with regards to either nutrient, then it would need to be listed on the label.

- The information on calories from fat will no longer be required.

- For sodium, dietary fiber, and vitamin D, updated daily values will be required on the label.

- The new rules will require listing the serving size that more closely reflect the amounts of food that people currently eat.

- Food packages with multiple serving which could be consumed in one or multiple sittings will require both "per serving" and "per package" calorie and nutrient information.

- Food packages that contain between one and two servings, the calories and other nutrients will be required to be labeled as one serving because it is typically consumed in one sitting.

- Larger type size for "Calories," "Servings per container," and the "Serving size" to make them more prominent.

- Bolding the "number of calories" and the "Serving size."

- An abbreviated footnote to better explain the %DV - "The % Daily Value tells you how much a nutrient in a serving of food contributes to a daily diet. 2,000 calories a day is used for general nutrition advice."

- The footnote table listing the reference values for certain nutrients for 2,000 and 2,500 calorie diets will not be required.

- The final rules become effective on July 26, 2016. The compliance date is July

26, 2018 for manufacturers with $10 million or more in annual food sales and July 26, 2019 for manufacturers with less than $10 million in annual food sales.

In December, 2014, the FDA issued final ruling on:

- 1) "Food labeling; nutrition labeling of standard menu items in restaurants and similar retail food establishments,"

- 2) "Food labeling; calorie labeling of articles of food in vending machines."

- Menu and Vending Machines Labeling Requirements – provides information on the requirements for menu and vending machine labeling, addresses frequently asked questions, and provides educational resources for the consumers.

- FDA has extended the compliance date to May 7, 2018 for menu labeling.

On August 2, 2013, FDA issued a final rule defining "gluten-free" for food labeling, which provides a uniform standard for manufacturers who choose to label their products as "gluten-free" and to help consumers, especially those living with celiac disease, be confident that items labeled "gluten-free" meet a defined standard for gluten content.

- Constituent Update - Foods Labeled Gluten-Free Must Now Meet FDA's Definition.

- Federal Register Notice for the Gluten-Free Labeling Final Rule.

- Questions & Answers: FDA's Final Rule on Gluten-Free Labeling.

- Proposed Rule for Gluten-Free Labeling of Fermented or Hydrolyzed Foods.

- Guidance for Industry: Gluten-Free Labeling of Foods; Small Entity Compliance Guide.

Other Food Label Regulations in the U.S.

The Federal Food, Drug and Cosmetic Act of 1938 (FD&C Act) gives FDA the authority to regulate food, inspect food plants, develop process to evaluate new ingredients, and prevent misbranding of food. Amendments to the FD&C Act related to food labeling include:

- Food Allergens Labeling and Consumer Protection Act of 2004 (FALCPA) requires that food that contains an ingredient that is an allergen or contains protein from a "major food allergen" should be declared on the food label. Eight foods or food groups are identified as major food allergens: milk, eggs, fish, crustacean shellfish, tree nuts, peanuts, wheat, and soybeans. FALCPA applies to both domestically manufactured and imported packaged foods that are subject to FDA regulation. USDA also applies FALCPA to products under its jurisdiction.

- ◦ FDA guidance for industry on allergen labeling. October, 2006.

- FDA Modernization Act of 1997 (FDAMA) authorizes health claims based on authoritative statements.

- Dietary Supplement Health and Education Act of 1994 (DSHEA) regulates labeling of dietary supplements.

Nutrition Labeling and Education Act of 1990 (NLEA) requires that nutrition information be made available on food packages to assist consumers in selecting foods that could lead to healthier diets, eliminate consumer confusion, and encourage industry innovation in product reformulations and health communications.

Label claims for conventional foods and dietary supplements – provides information on three categories (listed below) of claims permitted on food and dietary supplement labels.

- Health claims: describes relationship between a food substance (food, food component or dietary supplement ingredient) and reduced risk of a disease or health related condition. There are 3 categories of health claim:

  - ◦ Nutrition Labeling and Education Act of 1990 (NLEA) authorized health claim: permits the use of claims on labels that characterize a relationship between a food, food component or dietary supplement ingredient and risk of a disease.

  - ◦ Health claims based on authoritative statements: the FDAMA authorizes use of health claims based on an authoritative statement of the National Academy of Sciences or a scientific body of the U.S. government with responsibility for public health protection or nutrition research.

  - ◦ Qualified health claims: provides a mechanism to request FDA to review the scientific evidence when there is emerging evidence for a relationship between a food substance and reduced risk of a disease or health-related condition, but the evidence is not well enough established to meet the significant scientific agreement standard required for FDA to issue an authorized health claim regulation.

- Nutrient content claims: the NLEA permits the use of claims that characterize the level of a nutrient such as *free, high,* and *low,* or compare the level of a nutrient in a food to that of another food, using terms such as *more, reduced,* and *lite* in a food. Nutrient claims have to be authorized by the FDA.

- Structure/function claim: may describe the role of a nutrient or dietary ingredient intended to affect the normal structure or function of the human body, but cannot relate to the treatment or mitigation of a disease. Structure/function claims may appear on conventional foods and dietary supplements. Structure/ function claims are not pre-approved by the FDA.

The Food Labeling regulation in the Code of Federal Regulations Title 21 (Chapter 1, subchapter B, Part 101) provides information on:

- General labeling provisions
- Specific food labelling requirements
- Specific nutrition labeling requirements and guidelines
- Specific requirements for nutrient content claim
- Specific requirements for health claims
- Specific requirements for descriptive claims that are neither nutrient content claims nor health claims
- Exemptions from food labeling requirements

Labeling & Nutrition Guidance Documents & Regulatory Information – provides regulatory information and guidance documents for general, nutrition, and allergen labeling, and label claims.

- Guidance for Industry: A Food Labeling Guide – addresses frequently asked question related to food labeling.

The Federal Meat Inspection Act , Poultry Products Inspection Act, and Egg Products Inspection Act - prevents false labeling or misbranding of meat, poultry, and processed egg products.

- The Food Safety and Inspection Service (FSIS) of the USDA regulates meat, poultry and processed eggs products including labeling of these products.

- FSIS regulates products containing 3% or more by weight at formulation of raw meat or poultry; and 2% or more by weight at formulation of cooked meat or poultry.
  - A Guide to Federal Food Labeling Requirements for Meat, Poultry and Egg Products. August, 2007.

The Organic Foods Production Act promulgated in 1990, effective in 2000 is regulated by the USDA.

- The National Organic Program (NOP) – develops the regulations for the production, handling, labeling, trade, and compliance criteria for producing organic animals, plants, and foods therefrom, which are under USDA and FDA jurisdiction, and oversees the enforcement of NOP requirements administered by third-party accredited organic certifying entities.

- ○ Four categories of labeling claims are allowed:
- ○ 100% organic
- ○ Organic
- ○ Made with organic
- ○ X% organic ingredients

The Federal Trade Commission Act (FTC Act) – prohibits "unfair or deceptive acts or practices." In the case of food products, the FTC Act prohibit "any false advertisement" that is "misleading in a material respect." Since 1954, the FTC and the FDA have operated under a Memorandum of Understanding, under which the FTC has assumed primary responsibility for regulating food advertising, while FDA has taken primary responsibility for regulating food labeling.

- Enforcement Policy Statement on Food Advertising – provides guidance regarding FTC's enforcement policy with respect to the use of nutrient content and health claims in food advertising. May, 1994.

Importing Food Products into the U.S.:

- Under the FD&C Act, importers of food products into U.S. interstate commerce are responsible for ensuring that the products are safe, sanitary, and labeled according to U.S. requirements. Imported food products are subject to FDA inspection. Both imported and domestically-produced foods must meet the same legal requirements in the U.S.

- The FSIS is responsible for assuring that meat, poultry and egg products imported into the U.S. are safe, wholesome, unadulterated, and properly labeled and packaged.

Exporting Food Products from the U.S.: Food products exported from the U.S. should meet the packaging, labeling and other special conditions required by the country. Information related to food exports is available on the FDA and USDA website.

Food Label Regulations in other Countries

Australia and New Zealand

- Food Labelling – provides guidance documents on labelling requirements. August, 2013.

Brunei

- The Ministry of Health sets forth regulations on food labeling in their Public Health (Food) Act.

Canada

- Food and Drugs Act 1985 establishes regulatory authority over food. Last amended November, 2014.

- Canada's Regulatory System for Foods with Health Benefits - An Overview for Industry – provides an overview of Canada's food regulatory system and important resources on labelling.

China

- The Global Agricultural Information Network (GAIN) Report by the USDA Foreign Agricultural Service – provides information on basic principles and requirements for the nutrition labeling and claims on pre-packaged foods directly offered to consumers in China. December, 2013.

European Union

- Food Labelling Legislation – provides details on the new EU regulations which came into effect December 13, 2014. The requirement to provide nutrition information will apply from December 13, 2016. The new law combines the directives on labelling, presentation and advertising of foods, and nutrition labelling for foods into one legislation.

- The EU General Food Law Regulation (2002) lays down general principles and requirements of food law.

- The European Food Safety Authority evaluates nutrition and health claim petitions to ensure that the claim made on the food label is substantiated by scientific evidence and issues opinions.

India

- The Food Safety and Standards Authority of India regulates food labels. The Food Safety and Standards (Packaging and labelling) Regulations came into effect in 2011.

Japan

- The Consumer Affairs Agency established in 2009 is responsible for developing and introducing legislation regarding food labeling standards.

Malaysia

- The Ministry of Health regulates labeling of foods and beverages through authority vested in the Food Act of 1983 and the Food Regulations of 1985.

Mexico

- The GAIN Report by the USDA Foreign Agricultural Service – provides information on the labeling requirements for pre-packaged foods and non-alcoholic beverages commercialized in Mexico. April, 2010.

Philippines

- The 2014 "Revised rules and regulations governing the labeling of prepackaged food products," establishes labeling standards for packaged food products distributed in the Philippines.

Singapore

- The Agri-Food & Veterinary Authority of Singapore establishes Food Regulations including labeling requirements for all pre-packed food products sold in Singapore. Last amended August, 2013.

- A Guide to Food Labelling and Advertisements – helps food importers, manufacturers, and retailers better understand the labeling requirements of the Food Regulations. Last amended January, 2015.

- Labelling Guidelines for Food Importers & Manufacturers – provides an overview of labeling requirements for all pre-packed food products sold in Singapore.

Food Labeling Standards and Resources from Non-government organizations

Codex Alimentarius

- General Standard for the Labeling of Prepackaged Foods – these labeling standards apply to all prepackaged foods. Adopted 1985, last amended 2010.

- Guidelines for Use of Nutrition and Health Claims – relate to the use of nutrition and health claims in food labelling. Adopted 1997, last amended 2013.

- Guidelines on Nutrition Labeling – recommend procedures for the nutrition labelling of foods. Adopted 1985, last amended 2013.

- Codex Committee on Food Labeling and Codex Committee on Nutrition and Foods for Special Dietary Uses – develops guidelines for use of nutrition and health claims on food labels.

Food and Agriculture Organization of the United Nations

- Food labeling – provides information developed by governments and non-government international agencies on food labeling.

- Innovations in food labeling – this book illustrates the multiple purposes food labelling serves and the many steps needed to implement a successful labeling policy. FAO. 2010.

The Sustainability Issues of Food Packaging

Food packaging is aimed at protecting food between the stages of processing and consumption. After consumption, the food packaging needs to be disposed in a responsible manner. It is a major contributor to municipal solid waste (MSW). The use of plastic and non-biodegradable materials for packaging creates negative impact on the environment on disposal.

Advances in food processing and packaging play a primary role in keeping the United States food supply one of the safest in the world. Packaging protects food between processing and usage by the consumer. Following usage, food packaging must be removed in an environmentally responsible manner. Packaging technology must therefore balance food protection with other issues, including energy and material costs, heightened social and environmental consciousness, and strict regulations on pollutants and disposal of municipal solid waste.

MSW consists of items commonly thrown away, including packages, food scraps, yard trimmings, and durable items such as refrigerators and computers. Legislative and regulatory efforts to control packaging are based on the mistaken perception that packaging is the largest component of MSW. Instead, the Environmental Protection Agency found that only approximately 31% of the MSW generated in 2005 was from packaging-related materials; this percentage has remained relatively constant since the 1990s despite an increase in the total amount of MSW. Non-packaging sources such as newsprint, telephone books, and office communications generate more than twice as much MSW.

Nevertheless, food packaging is a noteworthy contributor to MSW because food is the only product class typically consumed three times per day by virtually every person. Accordingly, food packaging accounts for almost two-thirds of total packaging waste by volume. Moreover, food packaging is approximately 50% of total packaging sales.

Although the specific knowledge available has changed since publication of "Effective Management of Food Packaging: From Production to Disposal," the Institute of Food Technologists' first Scientific Status Summary on the relationship between food packaging and MSW, the issue remains poorly understood, complicating efforts to address the environmental impact of discarded packaging materials.

Consequently, IFT has issued a new Scientific Status Summary that describes the role of food packaging in the food supply chain, the types of materials used in food packaging, and the impact of food packaging on the environment. Appearing in the April 2007 issue of *Journal of Food Science*, the new Summary, "Food Packaging—Roles, Materials, and Environmental Issues," provides an overview of EPA's solid waste management guidelines and other waste management options, addresses disposal methods of and legislation on packaging disposal, and describes the current sustainable cradle-to-cradle concept, which replaces the cradle-to-grave emphasis. The sustainability goal of the cradle-to-cradle concept is to recover sufficient materials and energy in a way that imposes zero impact on future generations.

Food Packaging Roles and Materials

The principal roles of food packaging are to protect food products from outside influences and distribution damage, to contain the food, and to provide consumers with ingredient and nutrition information. Traceability, convenience, and tamper indication are secondary functions of increasing importance. The goal of food packaging is to contain food in a cost-effective way that satisfies industry requirements and consumer desires, maintains food safety, and minimizes environmental impact.

Package design and construction play significant roles in determining the shelf life of a food product. The right selection of packaging materials and technologies maintains product quality and freshness during distribution and storage. Materials that have traditionally been used in food packaging include glass, metals, paper and paperboards, and plastics. Today's food packages often combine several materials to exploit each material's functional or aesthetic properties.

As research to improve food packaging continues, advances in the field may affect the environmental impact of packaging.

Waste Management Approach

Proper waste management is important to protect human health and the environment and to preserve natural resources. EPA's guidelines for solid waste management emphasize the use of a hierarchical, integrated management approach involving source reduction, recycling, composting, combustion, and landfilling.

- Source Reduction (i.e., waste prevention) is reducing the amount and/or toxicity of the waste ultimately generated by changing the design, manufacture, purchase, or use of the original materials and products. EPA considers source reduction the best way to reduce the impact of solid waste on the environment because it avoids waste generation altogether.

  Source reduction encompasses using less packaging, designing products to last longer, and reusing products and materials. Specific ways to achieve source reduction include using thinner gauges of packaging materials, purchasing durable goods, purchasing larger sizes or refillable containers, and selecting nontoxic products.

- Recycling diverts materials from the waste stream to material recovery. Unlike reuse, which involves using a returned product in its original form, recycling involves reprocessing material into new products.

HOW IS ALUMINIUM RECYCLED?

1 — Aluminium is collected along with other metals from your recycling bin

2 — Recycling is collected and delivered to a special processing plant

3 — Using the latest technology all material streams are separated

Metal elements are melted and treated to make new raw materials

4 — The recycled aluminium is now used in the production of new products

5 — The aluminium is back on the shelf for you to buy in as little as sixty days

A typical recycling program entails collection, sorting and processing, manufacturing, and sale of recycled materials and products. Almost all packaging materials are technically recyclable, but economics favor easily identified materials such as glass, metal, high-density polyethylene, and polyethylene terephthalate.

- Composting, considered by EPA as a form of recycling, is the controlled aerobic or biological degradation of organic materials, such as food and yard wastes. Accordingly, it involves arranging organic materials into piles and providing sufficient moisture for aerobic decomposition by microorganisms. Because organic materials make up a large component of total MSW, composting is a valuable alternative to waste disposal.

- Combustion, the controlled burning of waste in a designated facility, is an increasingly attractive alternative for waste that cannot be recycled or composted. Reducing MSW volume by 70–90%, combustion incinerators can be equipped to produce steam that can either provide heat or generate electricity (waste-to-energy combustors). In 2004, the U.S. had 94 combustion facilities, 89 of which were waste-to-energy facilities, with a process capacity of approximately 95,000 tons/day or about 13% of MSW.

- Landfilling provides environmentally sound disposal of any remaining MSW and the residues of recycling and combustion operations. As waste disposal methods, both landfilling and combustion are governed by regulations issued under subtitle D of the Resource Conservation and Recovery Act. Thus, today's landfills are carefully designed structures in which waste is isolated from the surrounding environment and groundwater.

EPA also strives to motivate behavioral change in solid waste management through non-regulatory approaches such as pay-as-you-throw and *WasteWise*. In pay-as-you-throw systems, residents are charged for MSW services on the basis of the amount of trash they discard. This creates an incentive to generate less trash and increase material recovery through recycling and composting. *WasteWise* is a voluntary partnership between EPA and U.S. businesses, institutions, nonprofit organizations, and government agencies to prevent waste, promote recycling, and purchase products made from recycled contents.

Moreover, EPA's *Environmentally Preferable Purchasing* program helps federal agencies and other organizations purchase products with less effect on human health and the environment than other products that serve the same purpose. Pollution prevention is the primary focus, with a broader environmental scope than just waste reduction.

Disposal Statistics

The most recently compiled waste-generation statistics indicate that 245.7 million tons of MSW were generated in 2005, a decrease of 1.6 million tons since 2004. The decrease in waste generation is partly attributable to the decreased rate of individual waste generation.

EPA analyzes MSW in two ways:

1. By materials: paper and paperboards, glass, metals, plastics, rubber and leather, textiles, wood, food scraps, and yard trimmings.

2. By major product categories: containers and packaging (mainly waste from food packaging, such as soft drink cans, milk cartons, and cardboard boxes); nondurable goods (newspapers, magazines, books, office paper, tissues, and paper plates and cups); durable goods (household appliances, furniture and furnishings, carpets and rugs, rubber tires, batteries, and electronics); and other wastes.

The containers and packaging category remained relatively constant at about 31% of the total waste generated between 2003 and 2005. EPA analysis of individual MSW generation rate shows a relatively constant rate of 4.5 lb/person/day since the 1990s, excluding the years 2000 and 2004 when it reached an all time high of 4.6 lb/person/day.

Even though waste generation has steadily grown since 1960, recovery through recycling has also increased. In 2005, 79 million tons of MSW was recovered through recycling and composting— slightly more than 58.4 million tons by recycling and 20.6 million tons by composting. The net per capita recovery reached an all time high of 1.5 lb/person/day.

Recovery was the highest for the containers and packaging category, followed by nondurable goods.

Despite the trend of increased recovery rates, the quantity of MSW requiring disposal has historically risen as a result of the increase in amounts generated. In 2005, approximately 168 million tons of MSW was discarded into the municipal waste stream—33.4 million tons combusted prior to disposal and 133.3 million tons directly discarded in landfills. The total amount of MSW generated has actually declined slightly since 2004; however, it is too early to determine if this is a change in the overall trend or merely a small variation that will not be maintained.

Limitations of Solid Waste Management Practices

Proper waste management requires careful planning, financing, collection, and transportation. Solid waste generation increases with population expansion and economic development and poses several challenges. For example, source reduction and convenience are often opposing goals in food packaging. Convenience features such as unit packages, dispensability, and microwavability usually require additional packaging. Similarly, tamper-indication features also add to the amount of waste generated.

Moreover, recycling and reuse are influenced by the costs of transporting, sorting, and cleaning collected materials. Many recycled materials, primarily plastics and paperboard, are restricted from food-contact applications. And both combustion and landfilling can have negative impacts on the environment through release of greenhouse gases or contamination of air and groundwater.

These aspects of packaging design and disposal must be weighed against environmental concerns in packaging. Because consumers dictate what is produced by what they choose to buy, at some point they need to evaluate whether the convenience and added safety are worth the increase in materials.

Choosing Packaging Materials

The key to successful packaging is to select the package material and design that best satisfy competing needs with regard to product characteristics, marketing considerations, environmental and waste management issues, and cost. Balancing so many factors is difficult and also requires a different analysis for each product.

Factors to be considered include the properties of the packaging material, the type of food to be packaged, possible food/package interactions, the intended market for the product, and the desired product shelf life. Other factors include environmental conditions during storage and distribution, product end-use, eventual package disposal, and costs related to the package throughout the production and distribution process.

Ideally, a food package would consist of materials that maintain the quality and safety of the food over time; are attractive, convenient, and easy to use while conveying all the desired information; are made from renewable resources, thereby generating no waste for disposal; and are inexpensive. Rarely, if ever, do today's food packages meet these lofty goals. Creating a food package is as much art as science, trying to achieve the best overall result without falling below acceptable standards in any single category.

From a product characteristic perspective, the inertness and absolute barrier properties of glass make it the best material for most packaging applications. However, the economic and safety disadvantages of glass boost the use of alternatives such as plastics. While plastics offer a wide range of properties and are used in various food applications, their permeability is less than optimal—unlike metal, which is totally impervious to light, moisture, and air.

PACKAGING MATERIALS

Attempts to balance competing needs can sometimes be addressed by mixing packaging materials— such as combining different plastics through coextrusion or lamination—or by laminating plastics with foil or paper. Plain paper is not used to protect foods for long periods of time because it has poor barrier properties and is not heat sealable. When used as primary packaging, paper is almost always treated, coated, laminated, or impregnated with materials such as waxes, resins, or lacquers to improve functional and protective properties. In contrast, paperboard is seldom used for direct food contact, even though it is thicker than paper.

Ultimately, the consumer plays a significant role in package design. Consumer desires drive product sales, and the package is a significant sales tool. Although a bulk glass bottle might be the best material for fruit juice or a sports beverage, sales will be affected if competitors continue to use plastic to meet the consumer desire for a shatterproof, portable, single-serving container.

Minimizing Environmental Impact

The impact of packaging waste on the environment can be minimized by prudently selecting materials, following EPA guidelines, and reviewing expectations of packaging in terms of environmental impact. Still, the primary purpose of food packaging must continue to be maintaining the safety, wholesomeness, and quality of food. Knowledgeable efforts by industry, government, and consumers will promote continued improvement, and an understanding of the functional characteristics of packaging will prevent much of the wellintentioned but illadvised solutions that do not adequately account for both pre- and post-consumer packaging factors. New materials, combinations, and technologies will allow the move from cradle-to-grave to cradle-to-cradle by eliminating negative environmental impact altogether.

EPA Guidelines for Management of MSW

- *Mass-burn incinerators*: Mass-burn incinerators accept all types of as-is MSW except for items that are too large to go through the feed system. Integrated waste is placed on a grate that moves through the combustor while air is forced into the system above and below the grate to promote complete combustion. Mass-burn incinerators are distinct from other MWCs because they burn the waste in a single stationary chamber and are typically constructed on site. Most mass-burn facilities are installed with boilers to recover the combustion heat for energy production. In 2004, 65 of the total 89 WTE facilities in the United States employed mass-burn technology to process approximately 22 million tons of MSW.

- *Refuse-derived fuel incinerators:* Refuse-derived fuel incinerators use waste that has been preprocessed to remove noncombustibles and recyclables. The combustibles are shredded into a uniform fuel that has a higher heating value. An RDF facility may be equipped for only processing or combustion, or both. In 2004 half of the 20 RDF facilities in the United States did both processing and combustion while the remaining 10 were equally divided between processing only and combustion only. RDF incinerators had a capacity of 8 million tons of MSW in 2004.

- *Modular combustors*: As with mass-burn incinerators, modular combustors accept all waste without preprocessing but are typically smaller than mass burn. They are usually prefabricated off site and can be quickly assembled wherever they are needed. Modular combustors accounted for about 10% of the total U.S. MWC units in 2004.

Landfilling

Landfills provide environmentally sound disposal of any remaining MSW and the residues of recycling and combustion operations. The location and operation of landfills are governed by federal and state regulations, and today's landfills are carefully designed structures in which waste is isolated from the surrounding environment and groundwater. A properly designed MSW landfill manages leachate and collects landfill gases (methane and others) for potential use as an energy source. Having passed through or emerged from landfill waste, leachate contains soluble, suspended, or miscible materials from the waste. EPA is investigating a modification in landfill design known as a bioreactor that can enhance aerobic and/or anaerobic degradation of leachate and organic waste.

The growing awareness of environmental problems, including increased use of synthetic packaging materials coupled with slow degradation in landfills, has prompted the development of advanced landfill technology, environmental regulations for landfills, and biodegradable packaging materials. Modern landfills are well engineered to prevent environmental contamination and managed to ensure compliance with federal regulations or equivalent state regulations. EPA has established a landfill reclamation approach that enables expansion of existing MSW landfill capacity and preclusion of land acquisition for new landfills. EPA also runs the landfill methane outreach program, which is a voluntary program that promotes the use of landfill gas as a renewable energy source.

Having established biodegradation as a minor benefit in landfills, EPA has developed bioreactor landfills that are designed to rapidly degrade organic waste by adding liquid or air to speed microbial processes. There are 3 types of bioreactors: aerobic, anaerobic, and hybrid. An initiative by the EPA to identify bioreactor standards or recommend operating parameters is underway.

Other Disposal Methods

Anaerobic Degradation

The main form of degradation that occurs in landfills is anaerobic degradation or digestion. In anaerobic degradation or digestion, microorganisms slowly break down solid waste—primarily organic-based materials such as wood and paper—into primarily carbon dioxide, methane, and ammonia. Collecting and pumping leachate through the compacted solid waste can accelerate this process by inoculating the mass and providing a moisture source that promotes further degradation. To prevent groundwater contamination, leachate should be contained in a system, usually a combination of liners and storage systems. Ultimately, leachate is processed by a treatment facility to make a stable residue that can be disposed of safely. Anaerobic degradation is mostly used to treat biosolids and organic waste contaminants. More research is necessary to realize the full potential of anaerobic degradation in the management of solid waste.

Biodegradable Polymers

Biodegradable polymers are derived from replenishable agricultural feedstocks, animal sources, marine food processing industry wastes, or microbial sources. In addition to renewable raw ingredients, biodegradable materials break down to produce environmentally friendly products such as carbon dioxide, water, and quality compost.

Biodegradable polymers made from cellulose and starches have been in existence for decades, with the 1st exhibition of a cellulose-based polymer (which initiated the plastic industry) occurring in 1862. Cellophane is the most common cellulose-based biopolymer. Starch-based polymers, which swell and deform when exposed to moisture, include amylose, hydroxylpropylated starch, and dextrin. Other starch-based polymers are polylactide, polyhydroxyalkanoate (PHA), polyhydroxybuterate (PHB), and a copolymer of PHB and valeric acid (PHB/V). Made from lactic acid formed from microbial fermentation of starch derivatives, polylactide does not degrade when exposed to moisture. PHA, PHB, and PHB/V are also formed by bacterial action on starches. In addition, biodegradable films can also be produced from chitosan, which is derived from the chitin of crustacean and insect exoskeletons. Chitin is a biopolymer with a chemical structure similar to cellulose.

Edible films, thin layers of edible materials applied to food as a coating or placed on or between food components, are another form of biodegradable polymer. They serve several purposes, including inhibiting the migration of moisture, gases, and aromas and improving the food's mechanical integrity or handling characteristics. Edible films are derived from plant and animal sources such as zein, whey, collagen, and gelatin.

Synthetic polymers can also be made partially degradable by blending them with biopolymers, incorporating biodegradable components, or adding bioactive compounds.

The biocomponents are degraded to break the polymer into smaller components. Bioactive compounds work through various mechanisms. For example, they may be mixed with swelling agents, which expand the molecular structure of the plastic upon exposure to moisture to allow the bioactive compounds to break down the plastic.

Arguments supporting the development of biodegradable polymers range from addressing problems of solid waste disposal and litter to substituting renewable resources for nonrenewable resources as raw materials. Despite certain advantages, the use of biodegradable materials is not a solution to all solid waste management problems. A switch from synthetic polymers to biopolymers will have little impact on source reduction and incineration, but recycling could be complicated by the existence of blended or modified polymers unless they are separated from the recycling stream. Biodegradable plastics have little benefit in a landfill because landfills generally exclude the oxygen and moisture that are required for biodegradation. If biopolymers become widely used, it is questionable whether there will be sufficient plant materials to make sufficient quantities of packaging polymers and whether optimizing crops for such polymers will interfere with food production. At this time, bioplastics are more expensive than most petroleum-based polymers, so substitution would likely result in increased packaging cost.

Even if biodegradable packaging is not practical on a broad basis, the advantages are very significant for certain applications. The litter argument for biodegradable plastics has merit to the extent that biodegradable plastics will tend to break down and become less obtrusive after being littered. Biodegradability is important in the marine environment in which litter poses hazards to marine life. Biodegradability can also be useful in military applications for which traditional disposal options are lacking. Specific but minor functions for biodegradable polymers include limiting moisture, aroma, and lipid migration between food components.

Commercialization of bioplastics is under way. NatureWorks, LLC manufactures polylactide from natural products. After the original use, the polymer can be hydrolyzed to recover lactic acid, thereby approaching the cradle-to-cradle objective. In addition, Wal-Mart Inc. is using biopolymers by employing polylactide to package fresh cut produce.

Theoretically, all plastics require sorting, but the reality is that recycling is often restricted to easily identifiable polymers and systems, most notably high-density polyethylene milk bottles and PETE soda bottles. Other polymers can be comingled into thermoplastic resins used for items such as park benches and playground equipment, which decreases the pressure to sort by specific polymer. Because polylactide is destined for commercial composting, it requires its own code and mechanisms for sorting if this advantage is to be exploited.

Litter

Littering is the improper disposal of solid waste. Because beverage containers are a visible component in litter, 11 states have enacted bottle bills to help ensure a high rate of recycling or reuse and to reduce litter. These bottle bills or container deposit laws mandate a minimum refundable deposit on beer, soft drink, and other beverage containers, thereby providing an economic incentive to ensure the return of used bottles. Beverage containers made from metal, glass, and plastics have been the most notable recycling successes because they are easily identifiable and made of single materials that are recyclable. Alternatively, biodegradable packaging could slowly help remove unsightliness and the hazards to animal and marine life caused by litter. However, it is possible that the existence of biodegradable containers may cause people to be less careful with their discards, which could hamper recycling efforts.

Current Disposal Statistics

The most recently compiled waste generation statistics indicate that 245.7 million tons of MSW were generated in 2005, which is an increase of approximately 37% over the 179.6 million tons generated in 1988 and a decrease of 1.6 million tons from 2004 . The decrease in waste generation is attributed in part to the decrease in individual waste generation rate. EPA analyzes MSW in 2 ways: (1) by materials and (2) by major product categories.

In the product categories, containers and packaging represent mainly waste from food packaging such as soft drink cans, milk cartons, and cardboard boxes. Nondurable goods encompass newspapers, magazines, books, office paper, tissue, paper plates and cups, and clothing and footwear. Durable goods include household appliances, furniture and furnishings, carpets and rugs, rubber tires, batteries, and electronics.

MSW Generation Analysis

Table: Weight of products generateda in MSW, with detail on containers and packaging

Table shows the EPA breakdown of MSW by both materials and products generated in the municipal solid stream. Among materials, paper and paperboard accounted for 34.2% while food scraps accounted for 11.9% of the total MSW. Glass, aluminum, and plastics contributed 5.2%, 1.3% and 11.8%, respectively. In product categories, containers and packaging formed the highest portion of the total solid waste generated at 31.2% followed by nondurable goods at 25.9%.

The weight and percentage of products generated in municipal solid waste from 1960 to 2005 with details on containers and packaging are shown in Table 3 and 4. The general trend indicates a continued increase in overall tonnage generated over the years up to 2004 followed by a decline in 2005. The total amount of waste generated from containers and packaging showed an increasing trend since 1990 with a small decrease in 2003 and 2005 , but the percentage of total waste remained relatively constant at about 31% . Additionally, EPA analysis of individual MSW generation shows a relatively constant rate of 4.5 pounds per person per day since the 1990s with the exception of 2000 and 2004, when it was at an all-time high of 4.6 pounds per person.

Analysis of containers and packaging indicates that paper and paperboard were the single largest contributors with 39 million tons followed by plastics at 13.7 million tons and glass at 10.9 million tons. The tonnage for food scraps and plastic packaging has significantly increased since the 1990s.

MSW Recovery Analysis

While waste generation has grown quite steadily since 1960, recovery through recycling has also increased. In 2005, a total of 79 million tons of MSW were recovered through recycling and composting. Of this amount, slightly more than 58.4 million tons were recovered by recycling and the rest by composting. The net per capita recovery was at an all-time high of 1.5 pounds per person per day in 2005.

Table: Impact of packaging materials and recycling on MSW

Among the product categories, containers and packaging were the most recovered followed by nondurable goods. Table below shows the generation and recovery of materials in MSW for 2005. About 59% of paper and paperboard, 51% of metals, 25% of glass, and 9% of plastics generated in containers and packaging were recovered. Among the materials, recovery of yard trimmings was the highest at 62%, followed by paper and paperboard at 50% and metal at 37%.

In spite of the trend of increasing recovery rates, the quantity of MSW requiring disposal has historically risen due to the increase in amounts generated. In 2005 approximately 168 million tons of MSW were discarded into the municipal waste stream of which 33.4 million tons were combusted prior to disposal and 133.3 million tons were directly discarded in landfills. The total amount of MSW actually declined slightly from 2004; it is too soon to determine whether this is a change in the overall trend or merely a small variation that will not be maintained.

Proper waste management requires careful planning, financing, collection, and transportation. Solid waste generation increases with population expansion and economic development, which poses several challenges.

Source Reduction Compared to Convenience

Source reduction and convenience are often opposing pressures in food packaging. Convenience features such as unit packages, dispensability, and microwavability usually require additional packaging, which is directly at odds with source reduction efforts. Similarly, tamper indication features also add to the amount of waste generated. Consumers dictate what is produced by what they choose to buy, and industry will produce what consumers demand if it can be done profitably. At some point, consumers need to evaluate whether the convenience and added safety are worth the increase in materials. Source reduction can be accelerated if consumers are willing to accept the loss of convenience and modify their buying habits accordingly. Refillable plastic containers have been developed as a strategy for source reduction but their use has declined in favor of nonreturnable containers.

Two competing trends influence source reduction of packaging materials. One trend is toward more economical bulk packs that need less packaging material per unit of product. If the ratio of package dimensions remains constant, increased size will increase the enclosure dimensions as a square function and increase the volume as a cube function. Therefore, the volume increases more rapidly, resulting in less packaging per unit volume. The trend toward larger sizes therefore represents a source reduction. The competing trend is for convenience and portion servings, in which individual portions are packaged, thereby increasing packaging usage. If all of the food is consumed, unit packaging would increase MSW. However, large portion sizes for small families can lead to food waste and thus increase total discards.

Materials for reuse and recycling must be sufficiently cleaned to remove any safety hazard posed by contaminants. The materials are often washed with powerful detergents that create liquid waste that must be properly treated. Furthermore, transportation costs can be high, depending on the proximity of each plant. Shipment of reusable or recyclable containers over long distances may require more energy than is saved by refilling. Glass is a heavy material, and recycling crushed glass requires transportation of postconsumer glass to a limited number of glass manufacturing facilities. If oil prices increase, the transportation distance that can be justified decreases. Lifecycle analysis studies can help determine the environmental impacts and resource demands of different waste management scenarios.

An unintended negative consequence of bottle bills is the entry of potentially contaminated materials into a food environment when the beverage containers are brought in for redemption. For example, if a bottle were used for garden chemicals, gasoline transfer, or any other nonfood use prior to return, this contamination could pose a hazard at the place of return if it were a food establishment. Furthermore, if the bottle were not adequately cleaned before recycling, the contamination could ultimately transfer to a new package made with the recycled materials. Unless they are rinsed to remove food residues, used soft drink bottles can also attract insects and other pests into a food establishment and foster the growth of microorganisms. This concern exists among many food establishments. These potential problems can be resolved, but the costs subtract from the realized benefits. The use of recycling centers instead of food establishments reduces these concerns.

Landfilling Compared to the Environment

Because landfills have the potential to contaminate air and groundwater, proper design, construction, and management are essential to prevent environmental damage. Prior to 1970, landfills were sited on the most convenient, least expensive lands, such as wetlands, marshes, quarries, spent mines, and gravel pits. Environmental impact with regard to toxic matter generation was not considered. The only environmental consideration was to cover the solid waste with soil to reduce odors, litter, and rodents.

In 1991, the emergence of evidence that siting landfills in wetland areas created groundwater contamination caused the promulgation of MSW Landfills Criteria. The standards address location restrictions, operating practices, and requirements for composite liners, leachate collection and removal, and groundwater monitoring.

Improperly designed landfills contaminate groundwater when water from rain or the waste itself permeates the landfill and dissolves substances in the waste. Acidic/alkaline conditions can enhance the extraction of certain substances. Under the standards, composite liners prevent leachate from reaching groundwater and allow its collection and treatment before disposal. Even though these efforts minimize groundwater contamination, limiting air and water permeation of waste also hinders the degradation of organic material within landfills. EPA's research into bioreactors and support of composting are attempts to better address the management of organic waste.

Many MSW landfills are also subject to air emission standards. Landfill gas emissions contain methane, carbon dioxide, and more than 100 different nonmethane organic compounds such as vinyl chloride, toluene, and benzene. Air emission standards require gas collection and treatment systems; in addition, systems that incorporate energy recovery are encouraged.

Public opposition to siting of incinerators and landfills for waste disposal is described by the acronyms NIMBY, NIMED, and NIMTO. The siting problem is therefore not only an issue of technical significance but also economic, social, and political. Effective public involvement is a significant component of a comprehensive siting strategy.

Combustion Compared to the Environment

With the continued decline in landfill capacity, combustion—especially waste-to-energy combustion—is becoming a widely used method to address increased MSW disposal needs. However, with the exception of modular combustors, incinerators require considerable initial capital, and construction takes 3 to 5 y. In addition, incineration results in air emissions that must be considered and controlled. Carbon dioxide, a greenhouse gas, is released when products derived from fossil fuels are burned. Pollution concerns include the emission of particulate matter, acidic gases, heavy metals, halogens, dioxins, and products of incomplete combustion. Dioxins and halogens are released from incineration of chlorinated polymers, the most abundant of which is PVC, constituting approximately 1% of MSW. Incomplete combustion of the organic components of MSW is also possible with suboptimal operation of an incinerator.

Lead- and cadmium-based additives for plastics and colorants contribute to the heavy metal content of MWC ash. Although used in small quantity, these metals concentrate in the ash as the polymers are burned off. Ash disposal is currently managed as potentially hazardous material under Subtitle C of the Resource Conservation and Recovery Act. In addition, the Clean Air Act regulates MWCs. Several regulations are currently in place for new and existing MWCs. In 2004 nearly all MWCs were equipped with particulates and acid gas controls in compliance with state and federal standards.

Upcoming Trends

The food packaging industry has witnessed significant innovations in packaging materials in recent years. Modern food packaging strives to be higher, durable, sustainable, renewable and biobased. This chapter will explore the upcoming trends in food packaging such as Liqui Glide, Amcor's Liqui Form™ bottle, Atomic Layer Deposition (ALD), etc.

The food packaging industry is vibrant and highly competitive, with food manufacturers always on the look-out for packaging that can provide consumers with increased convenience as well as longer shelf life at a lower cost than their existing packaging. The food industry is well aware that consumers want innovation and value novelty, and therefore the packaging industry must innovate or stagnate. Given the size and diversity of the food packaging industry, this brief overview can only touch on a few of the major trends and innovations.

Material Substitution

LOOKS LIKE GLASS
BUT IT'S PET!

Over the past few decades there have been significant changes in the relative proportions of the packaging materials glass, metal, paper and plastics used to pack food. Most noticeable has been the switch from glass (and to a lesser extent metal) to plastics with, for example, the majority of beverages nowadays packed in polyethylene terephthalate (PET). It is just 40 years since the first PET bottle appeared in the market and its growth since then has been hugely successful. The last great glass markets under threat from PET are wine and beer and recently Sidel created a pasteurisable PET bottle that utilises a 'champagne' base traditionally found on glass beer bottles. It also supports a crown cap, which together with the non-petaloid base, gives the appearance of a glass bottle. A 330 mL bottle weighs only 28 g, which is up to 86% less than an average equivalent glass bottle; a 600 mL bottle is also available. The reported shelf life of up to 6 months is based on <1 ppm O_2 and <17% CO_2 loss which is a requirement of the brewing industry. Another big market for PET bottles is edible oil and recently a 3 L PET bottle with a handle was launched in the USA. Its light weight and convenience are likely to prove popular with consumers.

Lightweighting

Lightweighting has been going on for decades, driven primarily by economics but in recent years it has always been trumpeted as being driven by environmental concerns. Just when you think the limit has been reached, a new low is achieved. For example, Krones new PET Lite 9.9 bottle weighs just 9.9 g for a 500 mL carbonated beverage bottle and is 30 to 45% lighter than comparable PET containers on the market. Direct printing onto the bottle means that no label is required and a special neck finish enables a tearoff ring-pull closure to be attached. Another recent example is the Sidel Right Weight PET 500 mL bottle for water that weighs just 7.95 g. To put this into context, the industry average is 12 g and in 1985 the weight of a comparable bottle was 28 g. As well as being lighter, this latest bottle has 32% more top-load performance.

Smart Labels

The Universal Product Code is a bar code symbology used for scanning packages at point of sale. It has been widely used on food and other packs since its launch in 1974 on a 10-pack of chewing gum. Now a variety of bar code symbologies that can be read by smartphones are appearing on packs. The QR (quick response) is the most common – it can launch exclusive content, update your Facebook status, download coupons, promotions and music and invite your friends to join you. Guinness has unveiled a product-activated QR code printed on a glass that only becomes visible when the glass is full of dark beer. When the glass is empty, or filled with a pale amber beer, it cannot be seen.

Sustainability

Although sustainable packaging is widely discussed at conferences and in the packaging media, there is no consensus as to what it is. Many in the packaging industry are confused; consumers are also very confused and the potential exists for unscrupulous companies to market packages as 'sustainable' when they are not and thus mislead consumers. However, a single definition of sustainable packaging is unfeasible, as the sustainability of a packaging material intrinsically depends on aspects specific to its life cycle, such as its manufacturing process, the length of its supply chain, its use and finally its disposal options. Many professionals would even argue that there is no such thing as 'sustainable packaging'. Rather there are improvements that can be made to the packaging's attributes and its manufacturing process in order to reduce its life cycle impacts and improve the efficiency of the supply chain. This was confirmed in the conclusions to a 2012 report from PwC which found that sustainable packaging was no longer a relevant term today as it is too broad to be useful at a practical level. Furthermore, no one can come up with a single meaningful definition of sustainable packaging. As a consequence, sustainable packaging has been substituted with a more balanced view of efficient packaging: minimum resources, minimising product waste, transport and display efficiency and effective after-use disposal and recycling. UK-based INCPEN defines a sustainable packaging and product supply chain as a system that enables goods to be produced, distributed, used and recovered with minimum environmental impact at lowest social and economic cost.

Biobased but not Biodegradable Plastics

Sustainable means to maintain or keep going continuously and the word has been used in connection with forest management for over a century. To be sustainable, consumption of resources must match their rate of renewal and therefore the use of non-renewable resources, such as petroleumbased plastics (and metals), is unsustainable. This has led to a focus on renewable biobased plastics. Nature produces 170 billion metric tonnes per year of biomass by photosynthesis yet only 3-4% of this material is used by humans for food and non-food purposes. Biomass carbohydrates are the most abundant renewable resources available (75% of this biomass) and are currently viewed as a feedstock for the green chemistry of the future (including bioplastics).

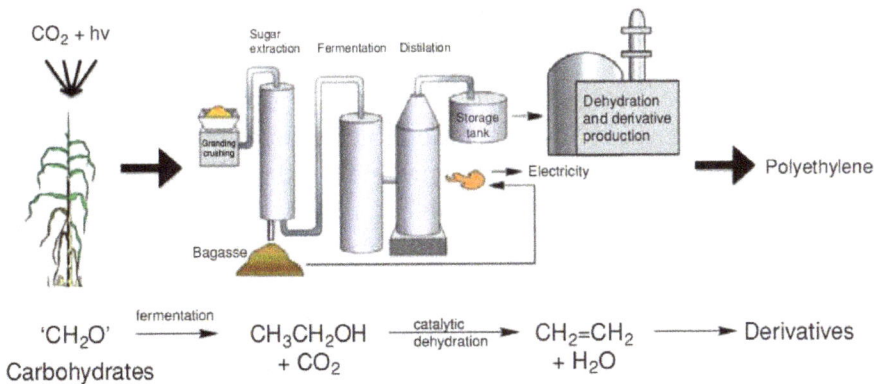

$$\text{'CH}_2\text{O'} \xrightarrow{\text{fermentation}} \text{CH}_3\text{CH}_2\text{OH} \xrightarrow[\text{dehydration}]{\text{catalytic}} \text{CH}_2\text{=CH}_2 \longrightarrow \text{Derivatives}$$

Carbohydrates · · · · · · · · · · · · · · · $+ \text{CO}_2$ · · · · · · · · · · · · · · · · $+ \text{H}_2\text{O}$

Figure: Flow diagram of the production of biopolyethylene from sugarcane via fermentation into ethanol and subsequent dehydration into ethylene

Bioethylene can be produced by the catalytic dehydration of bioethanol produced by the fermentation of carbohydrates, followed by normal polymerisation to produce poly-ethylene (PE) as shown in figure. It is not biodegradable and has the same properties, processing and performance as PE made from natural gas or oil feedstocks. The major producers are in Brazil and use sugar from cane as the starting material. Current applications by multinationals include yogurt cups (Danone), fruit juice bottles (Odwalla) and plastic caps and closures for aseptic paperboard cartons (Tetra Pak).

Such developments have led to the new paradigm for sustainable food packaging: biobased but not biodegradable. This is further evidenced by Coca-Cola's PET Plant-bottle where the ethylene glycol and the terephthalic acid are derived from plant-based sugars and agricultural residues. Although the 100% biobased bottle was re-leased in Milan in June 2015, it will be five to eight years before it is available in commercial quantities and it will not reach price parity (equalling the price of pro-ducing current Coca-Cola PET bottles) until 2018. Heinz and Danone will also have access to this bottle.

A very exciting development undertaken by Avantium in The Netherlands has result-ed in a new polyester: polyethylene furanoate (PEF), an analogue of PET. The main building block in PEF, 2,5-furandicarboxylic acid (FDCA), is derived from plant-based carbohydrates and can be used as a replacement for terephthalic acid. PEF could re-place PET in typical packaging applications, such as films and in particular bottles, as it outperforms PET in many areas, particularly barrier properties. Specifically, PEF's O_2 barrier is ten times better than that of PET, the CO_2 barrier is four times better and the H2O barrier is two times better. Pilot-scale production is currently underway. Of course sustainably managed forests have an assured future supplying paper and wood-based packaging materials. Earlier this year Carlsberg announced its ambition to develop the world's first fully biodegradable woodfibre (moulded pulp) bottle in conjunction with EcoXpac, which owns the rights to an energy saving, impulse-drying technique that it is claimed will 'disrupt the market for moulded fibres.' All parts of the bottle — including the cap — are to be manufactured using only biobased and biodegradable materials 'so they can be responsibly discarded and degraded.'

This raises the question as to why biodegradation is so popular among the general population and in the media. To many consumers biodegradation appears 'natural' – it is what nature does so it must be good! They believe biodegradable packaging will solve the solid waste problem plus the litter problem although few cities are able to collect and compost green waste. Biobased, biodegradable plastics have even been classed as 'sustainable packaging' by some people and organisations. But converting a solid material to a gas via biodegradation or composting cannot be sustainable. It is much better to recycle or recover the embodied energy through incineration. There is a need to close the resource loop and make the most out of the material rather than simply use it once.

Life Cycle Assessment (LCA) quantifies the resource and energy use as well as the environmental burdens over the entire life cycle of a package and is used to show how, for example, lightweighting and material changes lower environmental impacts. However, the conclusions cannot be extrapolated to provide universal generalisations as the results are specific to the precise system under study. There are a number of software packages available to perform LCAs and a recent study found that results from four LCA software systems disagreed on which package had the greatest environmental impact. Of real concern was the finding that some results were more than an order of magnitude different between software packages and discrepancies occurred in all four impact categories. In addition, all four software systems disagreed with each other at multiple points in the comparisons.

Pawelzik et al. reported that while internationally agreed LCA standards (ISO 14040 and 14044) provide generic recommendations on how to evaluate the environmental impacts of products and services, they do not address details that are specifically relevant for the life cycles of biobased materials. In particular, treatment of biogenic carbon storage is critical for quantifying GHG emissions of biobased materials in comparison with petroleum-based materials.

Innovations

Invention is the creation of a new idea, concept, device or process, while innovation is turning a new concept into commercial success — the introduction of change via something new. It follows that it is not an innovation until a customer says it is! In short, innovation = invention + exploitation. While the patent literature is full of inventions, few ever qualify as innovations. Drivers for packaging innovation include invention, fastchanging social trends, profitability, differentiation, environmental awareness and sustainability.

Among recent innovations that are finding application in the food packaging area are polymer-clay nanocomposites, plasma-enhanced chemical vapor deposition (used to deposit hydrocarbon films - sometimes referred to as amorphous carbon – on different substrates using, for example, acetylene in a plasma) and atomic layer deposition, all of which can improve the barrier properties of plastic and (in some cases) paper packaging. Space only permits a discussion of the latter.

Atomic Layer Deposition (ALD)

ALD was invented in 1974 by Dr Tuomo Suntola at the University of Helsinki. It is a surface-controlled, layer-by-layer, thin-film deposition process based on self-terminating gas-solid reactions. In packaging applications, metal oxides such as Al_2O_3, SiO_2 and ZnO are applied and a 10 nm oxide layer typically decreases the oxygen transmission rate by a factor of 10. Recently, water vapor transmission rate (WVTR) of ~6 x 10-3 g m-2 day-1 for PET using atmospheric ALD at a deposition temperature of 50°C was reported and when this invention is commercialized it will have a significant impact on food packaging structures.

LiquiGlide

Founded in 2012 from research at MIT, LiquiGlide coatings allow viscous liquids to move easily due to permanently wet slippery surfaces. The coatings consist of two layers: a porous solid layer and an impregnating liquid layer. The first consumer products with LiquiGlide coatings are expected to hit shelves in late 2015, with likely products including mayonnaise and ketchup. Benefits include reducing waste, increasing consumer value and eliminating the need for complicated pump modules and dispensing/closure systems.

Serac's Roll N Blow

Invented by Agami and commercialised by Serac, Roll N Blow uses an innovative tubular thermoforming technology to produce cups or bottles from plastic reels of polystyrene or polypropylene. Sizes range from 100 to 500 mL. The innovation lies in a vertical thermoforming process that first forms the plastic sheet into a pipe before heating and blowing the bottle into a mould. As a consequence, bottle designs are not limited to large necks and small heights and can be shaped totally round. Such bottles also show a better resistance to vertical compression than with flat thermoforming.

Amcor's LiquiForm Bottle

LiquiForm uses the consumable liquid instead of compressed air to hydraulically form and fill the PET container on one machine simultaneously. This simplifies the manufacture of rigid plastic containers and significantly reduces cost and waste. It differs from traditional blow moulding and filling operations in that when the preform is placed in the mould the actual beverage is forced at high pressure into the preform, moulding it into the bottle shape. This results in a filled bottle, ready for capping and labeling.

Some new items and fascinating trends on display are mentioned below.

Contrasting Ideas

Snack manufacturers have long put their savory and sweet treats in shiny film and foil bags and pouches. Recently, brands started trying out matte-finish packages to impart a smoother look. At this year's Sweets and Snacks Expo, multiple makers showed off packs with a mix of matte and spot-coated elements, offering an interesting contrast. Companies opting for this bold graphic effect include Mozaics Chips natural and organic snacks, and 4505 Meats pork cracklings.

Solo Snacking

During the first Tuesday morning session at the expo, Sally Lyons Wyatt, evp and practice leader of IRI, told attendees snacking tends to be a solitary experience—the average American snacks 2.5 times a day, and younger generations snack even more. Offering single-serve options (especially ones that intersect with an increased interest in healthy choices) are on the climb. Cookies are among these, including Cookie+ Protein and ProSupps' MyCookie.

Cake Bites

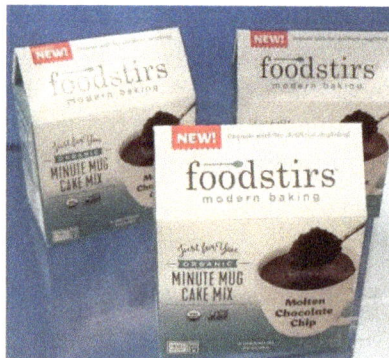

As the success of the meal prep kit business shows (with Blue Apron and Hello Fresh among the top stars), consumers often enjoy the do-it-yourself approach to eating, sometimes with a little head start. In that vein, mug cakes were found at a number of places around the show floor. Notable offerings include organic offerings from actress Sarah Michelle Gellar's powerhouse Foodstirs, as well as small-but-growing Molly & You.

Downsized Delectable

Sometimes a consumer just wants a little treat—but the snack they're craving, unfortunately, only comes in giant, mega-calorie versions. Increasingly, manufacturers are meeting this desire with pared-down versions of traditional treats, so that people can satisfy those cravings without totally busting their diets. Baked-good purveyor JJ's Bakery showed off its Tid-Bits mini snack pies at the show, roughly half the size of its traditional pies.

Multi-part Munchies

A number of candy and snack manufacturers showcased packages with multiple components, separating the ingredients until the consumer is ready to mix. The reasons for this division range from keeping crunchy parts from moisture, to delivering the delight of an interactive experience. Items on display that fell into this category include Purely Elizabeth's oatmeal (with crunchies in a separate compartment atop the cup) and MET-Rx Snack Cups (with a divider between high-protein snack sticks and hazelnut spread).

Same Candy, New Format

Putting an existing candy or snack into a new container can lead to innovation that connects with new consumers. Back in 2012, Kraft simply poured its crunchy mini-marshmallows into a shaker container used on its parmesan and created Mallow Bits. One example found on the show floor: Long Ball Licorice's Round Tripper, which puts traditional licorice in a multi-layer pack mimicking a chew pouch.

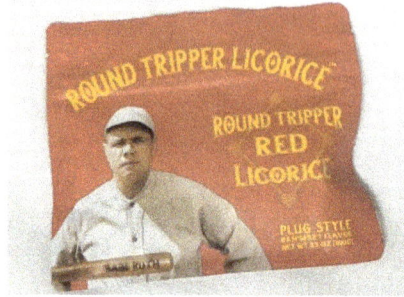

Fun with Branding

Tie-ins are nothing new—whenever a blockbuster action or comic-book movie is about to hit, the studio slaps its heroes on bags, boxes and jars on every shelf. What many new products at Sweets and Snacks showcased was a deeper, more fun approach to entertainment-branded candy and snacks more closely tying the pack promo and the product together. Examples at the show include Cosmos Creations' Dino Munch (dinosaur-shaped puffs promoting "Jurassic World: Fallen Kingdom") and Van Holten'sPickle Rick pouch, depicting a beloved "Rick & Morty" plotline.

Makeover Time

It's rare that a consumer packaged good never changes its look or package. Fun fact: The Swans Down flour box is a favorite of stage managers producing plays depicting decades gone by because it's stayed the same for so long. The makers of Smarties candy and Kar's Nuts snacks both unveiled makeovers of their longstanding logos and packs at the show.

Hip Chicks

Ahead of the event, NCA representatives predicted chickpeas would be a hot ingredient in many new products at this year's show. They were right on target—garbanzo beans starred in all sorts of products found in several Sweets and Snacks booths. Many were savory, like Biena's roasted chickpeas and Hippeas chickpea puffs—but interestingly, a number of companies are beginning to get bolder and put the humble legumes in sweet coatings, like chocolate.

Rice is Nice

Consumers increasingly care about their health and weight. Also, more people are looking to avoid ingredients like gluten, thanks to higher numbers of people getting celiac diagnoses. That could be one reason why a large number of manufacturers are coming out with rice-based edibles. Examples found at the show include Master Rice's R!se Buddy baked rice snacks and Crunchy Rollers snacks.

Permissions

All chapters in this book are published with permission under the Creative Commons Attribution Share Alike License or equivalent. Every chapter published in this book has been scrutinized by our experts. Their significance has been extensively debated. The topics covered herein carry significant information for a comprehensive understanding. They may even be implemented as practical applications or may be referred to as a beginning point for further studies.

We would like to thank the editorial team for lending their expertise to make the book truly unique. They have played a crucial role in the development of this book. Without their invaluable contributions this book wouldn't have been possible. They have made vital efforts to compile up to date information on the varied aspects of this subject to make this book a valuable addition to the collection of many professionals and students.

This book was conceptualized with the vision of imparting up-to-date and integrated information in this field. To ensure the same, a matchless editorial board was set up. Every individual on the board went through rigorous rounds of assessment to prove their worth. After which they invested a large part of their time researching and compiling the most relevant data for our readers.

The editorial board has been involved in producing this book since its inception. They have spent rigorous hours researching and exploring the diverse topics which have resulted in the successful publishing of this book. They have passed on their knowledge of decades through this book. To expedite this challenging task, the publisher supported the team at every step. A small team of assistant editors was also appointed to further simplify the editing procedure and attain best results for the readers.

Apart from the editorial board, the designing team has also invested a significant amount of their time in understanding the subject and creating the most relevant covers. They scrutinized every image to scout for the most suitable representation of the subject and create an appropriate cover for the book.

The publishing team has been an ardent support to the editorial, designing and production team. Their endless efforts to recruit the best for this project, has resulted in the accomplishment of this book. They are a veteran in the field of academics and their pool of knowledge is as vast as their experience in printing. Their expertise and guidance has proved useful at every step. Their uncompromising quality standards have made this book an exceptional effort. Their encouragement from time to time has been an inspiration for everyone.

The publisher and the editorial board hope that this book will prove to be a valuable piece of knowledge for students, practitioners and scholars across the globe.

Index

www.ingramcontent.com/pod-product-compliance
Lightning Source LLC
Chambersburg PA
CBHW062004190326
41458CB00009B/2959